ENGINEERING TE 10 MAN

10

IEE TELECOMMUNICATIONS SERIES 26

Series Editors: Professor J. E. Flood
Professor C. J. Hughes
Professor J. D. Parsons

COMMON-CHANNEL SIGNALLING

COMMON-CHANNEL SIGNALLING

Richard J Manterfield

Peter Peregrinus Ltd. on behalf of the Institution of Electrical Engineers

Published by: Peter Peregrinus Ltd., London, United Kingdom

Peter Peregrinus Ltd.,
Michael Faraday House,
Six Hills Way, Stevenage,
Herts. SG1 2AY, United Kingdom

British Library Cataloguing in Publication Data
Manterfield, R.
    Common-channel signalling.
    I. Title
    621.382

ISBN 0 86341 240 8

Printed in England by Short Run Press Ltd., Exeter

DEDICATION

To Liz, Muriel and Dorien.

Contents

7 Transaction capabilities

Preface

Signalling is the life-blood of telecommunications, transforming inert network elements into a powerful medium capable of providing services to customers. The modern trend is to introduce digital technology, particularly integrated services digital networks, and these conditions allow the full benefits of common-channel signalling (CCS) systems to be unleashed. CCS systems have a vast repertoire of signals and high speed of operation. These attributes, coupled with the derivation of signalling networks that are separate from other traffic, provide a major step in the evolutionary process towards unimpeded signalling-information transfer between customers, between customers and network nodes and within networks.

This book is aimed at a wide variety of readers. It is written to allow a novice to start with basic principles and build up a wide appreciation of CCS systems. Those readers who already have a knowledge of telecommunications, and wish to derive a better understanding of CCS systems, can select the more advanced chapters and avoid the basic concepts. The book is also intended to be used by experts in the CCS field as a reminder of terminology and concepts and to widen knowledge of CCS beyond a specific area.

The book starts with the principles of signalling systems in Chapter 1. Chapter 2 reviews the categories of channel- associated signalling system and Chapter 3 summarises CCITT Signalling System No.6. However, the focus is upon modern CCS systems and these are explained in Chapters 4 to 10. A fundamental attribute of modern CCS systems is evolutionary potential and their architecture, described in Chapter 4, is a key to their success. Chapters 5 to 7 describe the inter-nodal CCS system CCITT Signalling System No. 7. Chapters 8 and 9 describe the CCITT Digital Subscriber Signalling System No. 1. The interworking of CCS systems is explained in Chapter 10 and Chapter 11 reviews the book and draws some conclusions on how to meet customer needs in rapidly changing times.

Acknowledgements

I should like to thank members of British Telecommunications plc for their support in writing this book. Alan Misson, Chris Earnshaw, Keith Ward and Andy Valdar provided a boost to the project. Detailed comments were provided by Bryan Law, Peter Clarke, Cliff Wadsworth, Trevor Johnson, Garry Miller and Terry King: their efforts are much appreciated.

Thanks apply to Professor Flood for his painstaking task of reviewing my manuscripts. My thanks are also extended to the many participants at CCITT and other forums. It was both fun and hard work deriving recommendations and it was an honour to have worked with many experts from around the world for such a worthwhile cause. Some figures and text have been reproduced from relevant recommendations with prior authorisation by the International Telecommunications Union. The excerpts have been chosen to help explanation of principles, and responsibility for their choice is mine. The recommendations, forming the specifications of signalling systems, are available from the ITU, Place des Nations, Geneva, Switzerland.

Finally, thank you to all my friends and family for their support during the period of writing the book. Jackie and Karen made an invaluable contribution and I thank them for their support. Leo and Star helped in their inimitable style.

Abbreviations

ACM	Address-Complete Message
ACU	Acknowledgement-Signal Unit
CAS	Channel-Associated Signalling
CC	Connection Confirm
CCS	Common-Channel Signalling
CCITT	International Telegraph and Telephone Consultative Committee
CCITT No.6	CCITT Signalling System No.6
CCITT No.7	CCITT Signalling System No.7
CIC	Circuit-Identification Code
CLB	Clear Back
CR	Connection Request
CUG	Closed-User Group
CVT	Circuit-Validation Test
DDI	Direct-Dialling In
DPC	Destination-Point Code
DSS1	Digital Subscriber Signalling System No.1
DTID	Destination-Transaction Identity
DUP	Data User Part
FDM	Frequency-Division Multiplexing
FISU	Fill-in-Signal Unit
HDLC	High-Level Data-Link Control
IAM	Initial-Address Message
ISCP	ISDN-Signalling-Control Part
ISDN	Integrated Services Digital Network
ISO	International Standards Organisation
ISP	Intermediate-Service Part
ISUP	ISDN User Part
ITU	International Telecommunications Union
LI	Length Indicator
LSU	Lone-Signal Unit
LSSU	Link-Status Signal Unit
MBS	Multi-Block Synchronisation Signal Unit
MRVT	MTP Routeing-Verification Test
MSU	Message Signal Unit

MTP	Message-Transfer Part
MUM	Multi-Unit Message
NNG	National Number Group
NSDU	Network-Service Data Unit
NSP	Network-Service Part
OMAP	Operations, Maintenance and Administration Part
OPC	Originating-Point Code
OSI	Open-Systems Interconnection
OTID	Originating-Transaction Identity
PABX	Private Automatic Branch Exchange
PCM	Pulse-Code Modulation
PSPDN	Packet-Switched Public Data Network
REL	Release
RLC	Release Complete
RLG	Release Guard
RLSD	Released
RSN	Receive-Sequence Number
SABME	Set-Asynchronous-Balanced-Mode Extended
SAM	Subsequent-Address Message
SAP	Service-Access Point
SAPI	Service-Access-Point Identifier
SCCP	Signalling-Connection Control Part
SCU	System-Control Signal Unit
SDL	Specification and Description Language
SIF	Signalling-Information Field
SIO	Service-Information Octet
SLS	Signalling-Link Selection
SSN	Send-Sequence Number
SYU	Synchronisation-Signal Unit
TC	Transaction Capabilities
TCAP	Transaction-Capabilities Application Part
TDM	Time-Division Multiplexing
TEI	Terminal-Endpoint Identifier
TPIE	Transaction-Portion Information Element
TUP	Telephone-User Part
UI	Unnumbered Information
UP	User Part

Glossary

Term	Explanation	Context
ACCESS	The link between a customer and local exchange.	CCS
ACCESS SIGNALLING	Signalling between a customer and a local exchange.	CCS
ACCESS SIGNIFICANCE	Relevance of a message to the originating and terminating accesses.	CCS
ACCESS UNIT	The point at which access is gained to a packet-switched public data network.	DSS1
ACKNOWLEDGED OPERATION	A Layer 2 form of working in which each frame is numbered, thus permitting error control.	DSS1
ADDRESS-COMPLETE MESSAGE	A message used to indicate that sufficient address information has been supplied to reach the called customer.	No.6, No.7
ALERTING MESSAGE	A message indicating that a called customer is being notified of an incoming call.	DSS1
ANSWER MESSAGE	A message used to indicate that a called customer has answered a call.	No.6, No.7
ARCHITECTURE	The structured approach to the specification of CCS systems.	CCS
ASSOCIATED MODE OF OPERATION	The transfer of messages between network nodes over transmission links directly connecting the network nodes.	No.6, No.7
B CHANNEL	An access traffic channel.	DSS1

Term	Definition	Category
BACKWARD MESSAGE	A message sent towards the calling customer.	General
BASIC ACCESS	A means of connecting a customer to a local exchange in an ISDN using two traffic channels and a signalling channel.	DSS1
BASIC ERROR CORRECTION	A form of error-correction mechanism.	No.7 (MTP)
BEGIN MESSAGE	A message used to establish an association between two nodes.	No.7 (TC)
BROADCAST WORKING	The sending of signalling information to a range of customer terminals.	DSS1
BUFFER	A unit that stores CCS messages in preparation for transmission.	CCS
CALL FORWARDING	A supplementary service in which a call is diverted from one called customer to another.	No.7 (ISUP)
CALL-PROCEEDING MESSAGE	A message used to acknowledge a set-up message.	DSS1
CALL REARRANGEMENT PROCEDURE	A procedure that allows a customer to suspend a call, make changes to the terminal being used and then resume the call.	DSS1
CALL REFERENCE	A number that is used to identify a particular call.	CCS
CALLED - LINE IDENTIFICATION	The identity of the called customer.	No.7
CALLING-LINE IDENTIFICATION	The identity of the calling customer.	No.7
CALLING-PARTY RELEASE	A method of operation in which the calling customer controls, and usually releases, a call.	General
CCITT	An international organisation responsible for specifying standards for telecommunications.	General
CCITT SIGNALLING SYSTEM No.7	A common-channel-signalling system used between network nodes.	General

CHANGEBACK	A procedure for reversing changeover.	No.7 (MTP)
CHANGEOVER	A procedure for transferring signalling from one link to another.	No.7 (MTP)
CHANNEL-ASSOCIATED SIGNALLING (CAS)	A type of signalling in which signalling capacity is dedicated for use by a traffic circuit.	General
CHECK BITS	A field within a signal unit used for error-control purposes.	CCS
CIRCUIT-RELATED SIGNALLING	Signalling in which signals or messages are identified by the number of the traffic circuit to which the signals or messages refer.	General
CIRCUIT-SWITCHED DATA	Data using traffic circuits within transmission links in a similar way to telephone calls.	General
CIRCUIT-VALIDATION TEST (CVT)	An OMAP procedure that is used to check data consistency.	No.7 (TC)
CLEAR-BACK MESSAGE	A backward message used to initiate clearing of a speech circuit.	No.6, No.7 (TUP)
CLEAR-FORWARD MESSAGE	A forward message used to initiate clearing of a speech circuit.	No.6, No.7 (TUP)
CLOSED-USER GROUP	A supplementary service in which customers form groups or clubs.	General
CODESET	A group of information elements within which the codings are uniquely defined.	DSS1
COMMON-CHANNEL SIGNALLING (CCS)	A type of signalling in which signalling capacity is allocated from a common pool when required by a particular call.	General
COMPONENT	An element of information used to request a remote node to perform an action or to return the result of an action.	No.7 (TC)
COMPONENT SUB-LAYER	Part of TCAP responsible for requesting an action to be performed and returning the results of actions.	No.7 (TC)

CONFUSION MESSAGE	A message returned in response to receiving unrecognised information.	No.7 (ISUP)
CONNECT MESSAGE	A message denoting that a call is accepted by the called customer.	DSS1
CONNECT MESSAGE	A message combining the functions of address-complete and answer messages.	No.7 (ISUP)
CONNECTIONLESS	A type of information transfer that is accomplished without establishing a formal relationship between entities.	General
CONNECTION-CONFIRM (CC) MESSAGE	A message used to confirm the establishment of a signalling connection.	No.7 (SCCP)
CONNECTION-ORIENTED	A type of information transfer that is accomplished by establishing a formal relationship between entities.	General
CONNECTION-REQUEST (CR) MESSAGE	A message used to establish a signalling connection.	No.7 (SCCP)
CONSTRUCTOR	A complex type of information element that is structured in a recursive manner.	No.7 (TC)
CONTINUE MESSAGE	A message used to convey data between two nodes.	No.7 (TC)
D CHANNEL	An access signalling channel.	DSS1
DATABASE	A node storing information.	General
DATA-FORM MESSAGE	A message used to carry information in connection oriented signalling.	No.7 (SCCP)
DATA-LINK CONNECTION	A connection at Layer 2 between a local exchange and a customer terminal.	DSS1
DATA USER PART (DUP)	A user part defining procedures and formats for circuit-switched data.	No.7
DESTINATION-POINT CODE (DPC)	The point code of the node to which a message is destined.	No.7 (MTP)

DIALOGUE	The successive interchange of components between two nodes.	No.7 (TC)
DIGITAL CONNECTIVITY	A supplementary service in which a calling customer can request a call to be established using digital equipment.	No.7 (TUP)
DIGITAL-SUBSCRIBER-SIGNALLING SYSTEM No. 1 (DSS1)	A common-channel signalling system used between customers and ISDN exchanges.	General
DIRECT-DIALLING IN (DDI)	A feature allowing a call to be made to a specific extension connected to a private-automatic-branch exchange.	General
DISCONNECT FRAME	A frame used to clear a signalling connection between a customer and a local exchange in acknowledged operation.	DSS1
DISCONNECT MESSAGE	A message used to initiate the clearing of a call.	DSS1
DISCONNECTED CALL	A call in which a traffic circuit is cleared but not yet ready to be used for another call.	DSS1
DUAL SEIZURE	An attempt by two entities to select the same speech circuit.	General
DUAL SIGNIFICANCE	Relevance of a message to the originating access and the network or the terminating access and the network.	CCS
EN BLOC	A method of operation in which all address information is supplied by a calling customer in one batch.	General
END MESSAGE	A message used to finish an association between two nodes.	No.7 (TC)
END POINT	An entity representing the last point in a particular procedure.	General
END-TO-END SIGNALLING	Signalling between network nodes that is not analysed by intermediate nodes.	No.7
ENTITY	A network node or customer terminal.	General

ERROR CONTROL	A mechanism in CCS systems used to detect and correct errors in messages, e.g. due to corruption during transmission.	CCS
ERROR CORRECTION	The retransmission of messages in response to the detection of corruption.	CCS
ERROR DETECTION	The detection of errors (e.g. a corrupted message) in the transmission of messages.	CCS
FEATURE-KEY-MANAGEMENT PROTOCOL	A means of providing supplementary services by establishing a service profile for a customer.	DSS1
FILL-IN SIGNAL UNIT (FISU)	A signal unit used to maintain signalling-link alignment.	No.7
FIRST-PARTY RELEASE	A method of operation in which either the calling or called customer can release a call.	General
FIXED-MANDATORY FIELDS	Fields within messages that are compulsory and of fixed length.	CCS
FLAG	A field within a message used to de-limit the message.	CCS
FLAG	A bit used to indicate the source of a call reference value.	DSS1
FORMAT	The coding structure applicable to messages.	CCS
FORMAT CONSTITUENT	Part of the means of defining the interworking of CCS systems using formats.	CCS
FORWARD MESSAGE	A message sent towards the called customer.	General
FRAME-CHECK SEQUENCE	The check bits used by DSS1.	DSS1
FRAME IN A PCM SYSTEM	A collection of 8-bit codes relating to 24 or 30 traffic channels.	PCM
FRAME IN DSS1	A unit of signalling information.	DSS1

FREQUENCY-DIVISION MULTIPLEXING (FDM)	A means of assembling numerous traffic circuits, using the frequency spectrum, for transmission purposes.	General
FUNCTIONAL PROTOCOL	A means of providing supplementary services by relying on customer-terminal participation.	DSS1
GLOBAL SIGNIFICANCE	Relevance of a message to all portions of a call.	CCS
HEADING CODE	A field within a signal unit defining the class of message included in the signal unit.	CCS
HIGH-LEVEL DATA-LINK CONTROL	A protocol defined by the International Standards Organisation.	General
HOLD MESSAGE	A message used to reserve a traffic circuit and call reference during a call.	DSS1
I-FORMAT FRAME	A type of frame used to transfer information in acknowledged operation.	DSS1
IN-CALL MODIFICATION	A procedure allowing the modification of the characteristics of a speech circuit during a call.	No.7 (ISUP)
INCOMING SIGNALLING SYSTEM	A function, used in interworking of CCS systems, receiving a set-up or initial-address message.	CCS
INFORMATION ELEMENT	A field(s) within a DSS1 message.	DSS1
INFORMATION ELEMENT	The basic building block of TC messages.	No.7 (TC)
INITIAL-ADDRESS MESSAGE	A message used to initiate a circuit-related call in CCITT No.6 and CCITT No.7	No.6, No.7
INITIAL-SIGNAL UNIT	The first signal unit of a MUM in CCITT No.6	No.6
INTEGRATED SERVICES DIGITAL NETWORK (ISDN)	A network providing digital connections between customers.	General
INTER-EXCHANGE SIGNALLING	Signalling between exchanges within networks.	General

INTERMEDIATE-SERVICE PART (ISP)	Part of transaction capabilities comprising Layers 4 to 6 of the OSI model.	No.7 (TC)
INTER-NODAL SIGNALLING	Signalling between nodes within networks.	General
INTERNATIONAL STANDARDS ORGANISATION (ISO)	An international standards organisation (responsible for defining communications environments).	General
INVOKE	A component used to request an action to be performed.	No.7 (TC)
ISDN SIGNALLING-CONTROL PART	A form of ISUP in which call-control functions are separated from traffic-circuit-control functions.	No.7
ISDN USER PART (ISUP)	A user part defining procedures and formats for use in ISDNs.	No.7
KEYPAD PROTOCOL	A means of providing supplementary services by generating alpha-numeric codes.	DSS1
LABEL	A field within a signal unit defining the speech circuit to which the signal unit refers.	No.6, No.7
LAYER	A tier compliant with the OSI 7-Layer model.	General
LENGTH INDICATOR (LI)	A field giving the length of another field within a message.	CCS
LEVEL	A tier defined for use by CCITT No.7 circuit-related applications.	No.7
LINE SIGNALLING	Signalling dealing with setting up and clearing down traffic circuits.	General
LINK-STATUS SIGNAL UNIT (LSSU)	A signal unit used to indicate and monitor the status of a signalling link.	No.7
LOCAL SIGNIFICANCE	Relevance of a message to either the originating or the terminating access.	CCS
LOOK AHEAD	A procedure in which a node checks the availability of a called customer before establishing a speech circuit.	No.7 (TC)

LONE-SIGNAL UNIT (LSU)	A type of signal unit in CCITT No.6 that contains a whole message.	No.6
LOOP-DISCONNECT SIGNALLING SYSTEM	A signalling system in which signals are transferred by modifying the status of a direct-current loop.	General
MANAGEMENT INHIBIT	A procedure facilitating maintenance or testing.	No.7 (MTP)
MESSAGE	A unit of information in common-channel signalling systems.	CCS
MESSAGE SIGNAL UNIT (MSU)	A signal unit used to carry user-part information.	No.7
MESSAGE-TRANSFER PART (MTP)	Levels 1 to 3 of CCITT No.7, responsible for successfully transferring messages from one node to another, even under failure conditions.	No.7
MODE OF OPERATION	The means of routeing messages through a signalling network.	No.6, No.7
MTP ROUTEING-VERIFICATION TEST (MRVT)	An OMAP procedure that is used to verify MTP routeing data.	No.7 (TC)
MULTI-FRAME	A collection of 12 or 16 frames in PCM systems.	PCM
MULTI-FREQUENCY INTER-REGISTER SIGNALLING SYSTEM	A signalling system using compounded frequencies (2 out N frequencies) to transfer signals.	General
MULTI-UNIT MESSAGE (MUM)	A type of signal unit in CCITT No.6.	No.6
NATIONAL NUMBER-GROUP CODE	The number that is normally associated with a geographic area when dialling a trunk call.	General
NETWORK-SERVICE DATA UNIT	Blocks of data in non-circuit-related applications.	No.7 (SCCP)
NETWORK-SERVICE PART	The combination of the SCCP and MTP.	No.7
NETWORK SIGNALLING	Signalling to help maintain, operate and administer networks.	General

NODE	A basic element of a network, e.g. an exchange.	General
NON-ASSOCIATED MODE OF OPERATION	A means of transferring messages between network nodes over transmission links that do not directly connect the network nodes.	No.6, No.7
NON-CIRCUIT-RELATED SIGNALLING	Signalling in which messages are identified by a reference number, thus allowing messages to be transferred in the absence of traffic circuits.	General
OFF-LINE APPLICATIONS	Actions that do not need to be performed in a short time-scale.	General
OPEN-SYSTEM INTER-CONNECTION	A method of providing generic interfaces, allowing different equipment types to interwork effectively.	General
OPERATION	An action that one node requests another node to perform.	No.7 (TC)
OPERATIONS, MAINTENANCE AND ADMINISTRATION PART (OMAP)	A user of TC designed to control a signalling network.	No.7 (TC)
OPTIONAL FIELDS	Fields within messages that are not compulsory.	CCS
ORIGINATING-POINT CODE	The point code of the node sending a message.	No.7 (MTP)
OUTBAND SIGNALLING SYSTEM	A signalling system in which signals are transferred in frequency spectra additional to those employed by traffic circuits.	General
OUTGOING SIGNALLING SYSTEM	A function, used in interworking of CCS systems, sending a set-up or initial-address message.	CCS
OVERLAP	A method of operation in which address information is supplied by a calling customer in more than one batch.	General
PACKET DATA	Data divided into blocks, each block being routed independently through the network.	General

PACKET HANDLER	Equipment within exchanges capable of handling packet data.	General
PARAMETERS	Information contained in fields within messages.	CCS
PASS ALONG	A form of end-to-end signalling.	No.7 (ISUP)
POINT CODE	The unique identity of a signalling point.	No.7 (MTP)
POINT-TO-POINT WORKING	Exchange of signalling between a local exchange and a specific customer terminal.	DSS1
POINTER	A field used to indicate the location of another field within a message.	CCS
POST-DIALLING DELAY	The time between a calling customer completing dialling and receiving a tone or announcement (e.g. ring tone).	General
PRE-ARRANGED END	A procedure in which a dialogue is deemed to have ended unless a positive action is taken to signify otherwise.	No.7 (TC)
PREVENTIVE-CYCLIC RETRANSMISSION ERROR CORRECTION	A form of error-correction mechanism.	No.7 (MTP)
PRIMARY ACCESS	A means of connecting a customer to a local exchange in an ISDN using 30 traffic channels and a signalling channel.	General
PRIMITIVE (OSI)	A unit of information passed between adjacent tiers within an entity.	General
PRIMITIVE	A simple type of information element.	No.7 (TC)
PRIMITIVE CONSTITUENT	Part of the means of defining the interworking of CCS systems using OSI primitives.	CCS
PROCEDURE CONSTITUENT	Part of the means of defining the interworking of CCS systems using procedures.	CCS

PROCEDURES	The logical sequence of events that can occur during calls.	General
PROTOCOL	The combination of primitives, procedures and formats applicable to a tier.	CCS
PROTOCOL CLASSES	Types of service offered by the SCCP.	No.7 (SCCP)
PULSE-CODE MODULATION (PCM)	A method of converting information from an analogue form to a digital form for transmission over digital transmission systems.	PCM
QUASI-ASSOCIATED MODE OF OPERATION	A form of the non-associated mode of operation in which the network pre-determines the routeing of messages.	No.6, No.7
REAL-TIME APPLICATIONS	Actions that need to be performed in a short time-scale, e.g. during call establishment.	General
REASSEMBLY	The recombination of small blocks of data.	General
REDIRECTION	A supplementary service in which a call is diverted from one called customer to another.	No.7 (TUP)
REFERENCE NUMBER	A number used to identify a particular transaction.	General
REJECT	A component used to signify unrecognised information.	No.7 (TC)
RELAY POINT	A node relaying SCCP messages between other nodes.	No.7 (SCCP)
RELEASE (REL) MESSAGE	A message used in the clearing of a call.	DSS1, No.7 (ISUP)
RELEASE-COMPLETE (RLC) MESSAGE	A message used to confirm the end of clear-down procedures.	DSS1, No.7
RELEASE-GUARD MESSAGE	A message used to confirm that a speech circuit has been cleared.	No.6, No.7 (TUP)
RELEASED CALL	A call in which a traffic circuit is cleared-down and ready for use on another call.	DSS1

RELEASED (RLSD) MESSAGE	A message used in clear-down procedures.	DSS1, No.7
RESET	A procedure to overcome an uncertain status of a speech circuit.	No.7
RESTART PROCEDURE	A procedure in which traffic circuits are returned to an idle condition.	DSS1
RESUME	A feature allowing cessation of the suspend feature.	No.7 (ISUP)
RETRIEVE MESSAGE	A message used to reverse a hold message.	DSS1
RETURN ERROR	A component used to indicate that an action has not been performed.	No.7 (TC)
RETURN RESULT	A component used to supply the results of performing an action.	No.7 (TC)
ROUTEING LABEL	A field identifying the originating and destination nodes of a message.	No.7
SCCP METHOD	A form of end-to-end signalling.	No.7 (ISUP)
SIGNALLING CONNECTION	A signalling relationship established between entities.	General
SEGMENTATION	The splitting of data into small blocks.	General
SEIZE	The act of reserving a resource, e.g. reserving a speech circuit for a telephone call.	General
SELECTION SIGNALLING	Signalling to transfer address digits.	General
SEQUENCE NUMBERS	Numbers used in error-control mechanisms to detect loss of messages.	CCS
SERVICE	The functions performed by a number of layers for use by higher layers.	General
SERVICE-ACCESS POINT (SAP)	The point at which Layer 2 services are offered to Layer 3.	DSS1
SERVICE-INFORMATION OCTET	A field defining the appropriate user part.	No.7 (MTP)

SET-ASYNCHRONOUS-BALANCED-MODE-EXTENDED FRAME	A frame used to establish a signalling connection between a customer and a local exchange in acknowledged operation.	DSS1
SET-UP MESSAGE	A message used to initiate establishment of a call.	DSS1
S-FORMAT FRAME	A type of frame used in the operation of an access signalling channel.	DSS1
SHIFT PROCEDURE	A means of changing the value of the currently used codeset.	DSS1
SIGNAL-TRANSFER POINT	A network node that routes messages from one signalling point to another using Levels 1 to 3 of CCITT No.7.	No.7
SIGNAL UNIT	A block of signalling information.	No.7
SIGNALLING	The transfer of information between customers and networks, within networks and between customers.	General
SIGNALLING CONNECTION	A signalling relationship established between entities.	General
SIGNALLING-CONNECTION-CONTROL PART (SCCP)	Used, in conjunction with the message-transfer part of CCITT No.7, to provide a network service (i.e. Layers 1 to 3 of the OSI 7-Layer model).	No.7
SIGNALLING-INFORMATION FIELD (SIF)	A field containing user-part information.	No.7 (MTP)
SIGNALLING LINK	The direct interconnection of two signalling points for signalling purposes.	No.7
SIGNALLING-LINK ACTIVATION	A procedure preparing a signalling link for service.	No.7 (MTP)
SIGNALLING-LINK MANAGEMENT	A form of signalling-network management controlling signalling links.	No.7 (MTP)
SIGNALLING-LINK RESTORATION	A procedure restoring a signalling link into service after a failure.	No.7 (MTP)

SIGNALLING-MESSAGE HANDLING	Level 3 functions covering the routeing of messages through the network.	No.7 (MTP)
SIGNALLING-NETWORK FUNCTIONS	The Level 3 functions of the MTP.	No.7 (MTP)
SIGNALLING-NETWORK MANAGEMENT	Level 3 functions controlling the signalling network.	No.7 (MTP)
SIGNALLING POINT	A network node capable of operating CCITT No.7.	No.7
SIGNALLING RELATION	An ability for two network nodes to exchange signalling information.	General
SIGNALLING ROUTE	A collection of signalling links.	No.7 (MTP)
SIGNALLING-ROUTE MANAGEMENT	A form of signalling-network management covering the distribution of information on the signalling-network status.	No.7 (MTP)
SIGNALLING-TRAFFIC FLOW CONTROL	A means of limiting signalling traffic.	CCS
SIGNALLING-TRAFFIC MANAGEMENT	A form of signalling-network management covering the reconfiguration of signalling traffic.	No.7 (MTP)
SOURCE-LOCAL REFERENCE	The identity of the node originating an SCCP connection.	No.7 (SCCP)
SPECIFICATION AND DESCRIPTION LANGUAGE	A method of defining the procedures used in systems.	General
STRUCTURED DIALOGUE	A dialogue in which an explicit association between two nodes is established.	No.7 (TC)
SUBSEQUENT-ADDRESS MESSAGE (SAM)	Message used to provide address information additional to that provided in an IAM.	No.6, No.7
SUPPLEMENTARY SERVICES	Features and facilities additional to a basic service.	General
SUSPEND	A feature allowing the calling and called customers to suspend communication temporarily.	No.7 (ISUP)

TELEPHONE USER PART (TUP)	A user part defining procedures and formats for telephony.	No.7
TERMINAL END-POINT IDENTIFIER (TEI)	A number identifying a specific customer terminal.	DSS1
TIER	A self-contained group of functions, forming part of the architecture of a signalling system.	CCS
TIME-DIVISION MULTIPLEXING (TDM)	A means of assembling numerous traffic circuits, using time-slots, for transmission purposes.	General
TRANSACTION CAPABILITIES (TC)	A generic protocol used to transfer non-circuit-related information between network nodes.	No.7
TRANSACTION-CAPABILITIES APPLICATION PART (TCAP)	Part of TC within Layer 7 of the OSI model.	No.7 (TC)
TRANSACTION IDENTITY	A number used to identify to which dialogue a message pertains.	No.7 (TC)
TRANSACTION SUB-LAYER	Part of TCAP responsible for establishing and maintaining a connection between nodes.	No.7 (TC)
TRANSFER ALLOWED	A procedure reversing transfer prohibited.	No.7 (MTP)
TRANSFER PROHIBITED	A procedure preventing access to a signal-transfer point.	No.7 (MTP)
UNACKNOWLEDGED OPERATION	A Layer 2 form of working in which frames are not numbered, thus obviating the provision of error control.	DSS1
UNIDIRECTIONAL MESSAGE	A message used in TC in an unstructured dialogue.	No.7 (TC)
UNITDATA MESSAGE	A message used to carry information in connectionless signalling.	No.7 (SCCP)
U-FORMAT FRAME	A type of frame used to transfer information in unacknowledged operation.	DSS1
UNSTRUCTURED DIALOGUE	A dialogue in which an explicit association between two nodes is not established.	No.7 (TC)

USER	A customer or a function within a network node that makes use of a protocol.	General
USER PART (UP)	Level 4 of CCITT No.7, defining the meaning of messages and the logical sequence of events that occur during calls.	No.7
USER-TO-USER SIGNALLING	Signalling between customers that is not analysed by network nodes.	DSS1, No.7
VARIABLE-MANDATORY FIELDS	Fields within messages that are compulsory and of variable length.	CCS
VOICE-FREQUENCY SIGNALLING SYSTEM	A signalling system in which signals are transferred as tones within traffic circuits.	General

Chapter 1

Principles of signalling systems

1.1 Introduction

Telecommunications is the means by which a customer communicates with others using electronic signals. Transmission links are required to allow customers to communicate with each other. These links can consist of metal cables (e.g. copper cables), optical fibres or radio links (e.g. satellite links). If each customer wished to communicate with only a few others, it would be possible to provide a permanent transmission link for each pair of customers. However, generally, each customer wishes to communicate with a large number of others and the provision of permanent customer-to-customer transmission links is untenable. Hence, telecommunications exchanges are provided to allow customers to share the use of transmission links. The combination of exchanges and transmission links forms the basis of telecommunications networks.

Each transmission link in a network provides a number of circuits, as shown in Fig. 1.1. For example, if two optical fibres are provided between

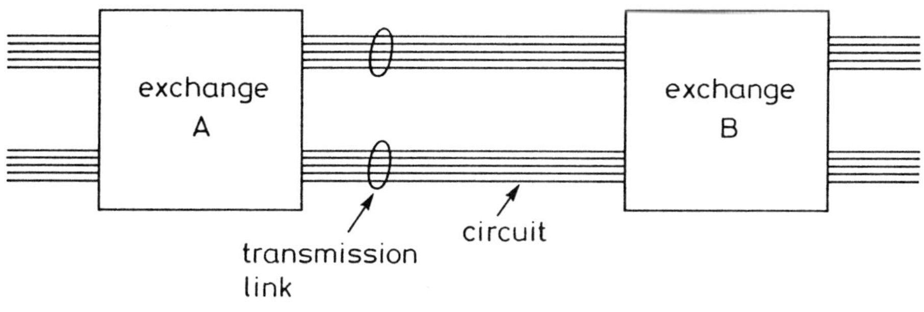

Fig. 1.1   Circuits within transmission links

Exchanges A and B, then each optical fibre can be divided, or multiplexed, into numerous circuits (typically 64 kbit/s circuits). Each of these circuits can provide a path for telecommunications traffic (speech or data) between the two exchanges. The job of an exchange is to connect one circuit with another (i.e. provide a switching function). By providing such connections on demand, and releasing them when the communication between two customers is completed, the circuits within transmission links can be re-used by other customers.

Telecommunications networks are usually built in a structured manner, a typical form being shown in Fig. 1.2. Customers are connected to the network by dedicated transmission links, typically to a local exchange. Each

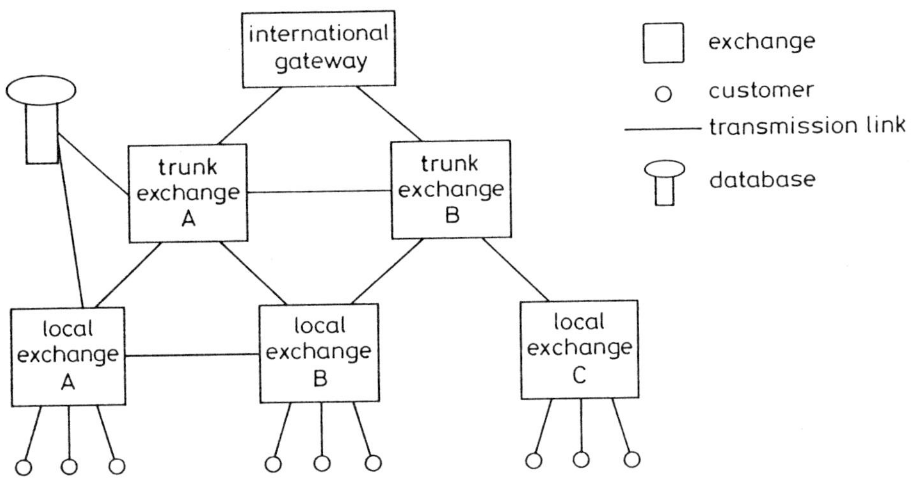

Fig. 1.2   Typical network structure

local exchange is connected to at least one trunk exchange and possibly other local exchanges. In these connections, the transmission links are shared by all customers. Trunk exchanges can be connected to an international gateway exchange, providing access to other countries. Telephone and data networks are extremely complex, both in terms of the technology needed to provide services and the operational support that is necessary to maintain effective working. As telecommunications services become even more sophisticated, additional network elements are required to support the functions of exchanges, e.g. databases can be provided to store specialised information. The collective term 'nodes' is used to describe exchanges and such additional network elements.

So, what is signalling? Signalling provides the ability to transfer information between customers, within networks and between customers and networks. Signalling is the life-blood, the vitalising influence, of telecommunications networks. It provides the bond that holds together the multitude of transmission links and nodes in a network to provide a cohesive entity. Without signalling, networks are inert. By providing effective signalling systems, a network is transformed into a tremendously powerful medium through which customers can communicate with each other using a range of telecommunications services. The critical nature of signalling is driving its rapid evolution. Old signalling systems that were simple mechanisms for transferring basic information are being replaced by efficient data-transfer highways. The ultimate objective is to provide an unimpeded transfer of information between customers, between nodes within networks (inter-nodal or inter-exchange signalling) and between customers and exchanges (access signalling).

Consider a customer connected to Local Exchange A in Fig. 1.2 who wishes to communicate with a customer connected to Local Exchange C. A speech path can be established from Local Exchange A to Trunk Exchange A over a circuit within the transmission link connecting the two exchanges. Similarly , a speech path can be established to Trunk Exchange B and subsequently to Local Exchange C, each exchange connecting appropriate circuits. Hence, a call (either a telephone call or a data call) can be established between the two customers and communication can take place. Upon completion of the communication, the call can be released and the circuits that were used in the various transmission links can be relinquished and made available for other customers to use. When the call is being established, the call is described as being in the 'set-up' phase. Having been established, the call is described as being in the 'conversation/data' phase. During the period in which the call is being released, the call is described as being in the 'release' or 'clear-down' phase.

In this example, it is signalling that provides the ability for the customer connected to Exchange A to indicate to the exchange that a call is required. It is signalling that allows the called customer to be identified, e.g. by transferring the telephone number dialled by the calling customer. It is signalling that allows the transfer of information between exchanges in the network to establish and release the call. It is signalling that transforms the foundation of the network into an active entity that can provide the required service to the customer.

Signalling information can be transferred between exchanges in two ways. Consider two exchanges in a telecommunications network that are connected by transmission links. Assume that the exchanges need to transfer information, e.g. to establish a speech path over a circuit within a transmission link. Two basic types of signalling system can be used. The first type of signalling system is termed 'channel-associated'. In such a system, a dedicated means of transferring signalling information is provided for each circuit. In channel-associated signalling (CAS) systems, the signalling information is either transferred over the circuit itself or dedicated signalling capacity is provided for each circuit within the transmission link. The second type of signalling system is termed 'common-channel'. In a system of this type, a common-signalling path is provided for a number of circuits. Hence, in common-channel signalling (CCS) systems, signalling capacity is provided in a common pool and allocated for use by each circuit as and when required.

The same principles apply to signalling systems used between customers and local exchanges (access signalling). In CAS access systems, the signalling information is usually transferred within the circuit itself. In CCS access systems, a common pool of capacity is provided to transfer information relating to a number of circuits.

The term 'channel', if applied strictly, refers to one-way transmission over a circuit. Thus, a circuit comprises two channels to allow two-way communication to take place. However, the terms are not always applied strictly and this can lead to confusion over which term to use. This book generally uses the term 'circuit' to refer to traffic paths and the term 'channel' to refer to CCS system paths. However, the term 'speech channel' is used in some chapters when this terminology is in common use within a specific context. This confusing terminology does not affect the principles of the signalling systems described.

1.2 Scope of book

In general terms, CAS systems are designed for use in 'old-technology' networks, in which exchanges use analogue techniques and transmission systems are primarily analogue (although some digital transmission systems are used between exchanges). Modern-CCS systems, on the other hand, are optimised for 'modern-technology' networks, in which both exchanges and transmission systems adopt digital techniques. Most national telecommunications networks still contain a great deal of equipment using CAS systems. Even when modern-technology exchanges are introduced, interworking with CAS systems is required. Hence, a description of the principles involved with CAS systems is given in Section 1.3 and an outline of the categories of CAS system is given in Chapter 2. Further details of CAS systems are well documented [1]. However, the main drive in modern networks is towards CCS systems and this book concentrates on these systems. The principles of CCS systems are covered in Section 1.4. The International Telegraph and Telephone Consultative Committee (CCITT), part of the International Telecommunications Union (ITU), has defined three CCS systems. CCITT Signalling System No 6 was defined during the early development of CCS and it is described in Chapter 3. Details of the two modern CCS systems, CCITT Signalling System No 7 (CCITT No. 7) and Digital Subscriber Signalling System No. 1 (DSS1), are given in Chapters 4 to

10. CCITT No. 7 is designed as an inter-nodal signalling system and DSS1 is designed as an access signalling system.

CCITT defines signalling systems in a series of 'recommendations'. The objective is for all member countries of CCITT to agree specifications that can be built by telecommunications-equipment manufacturers. International specifications of this nature reflect the ideas and views of experts throughout the world and allow customers in all countries to benefit from the provision of standards, resulting in economies of scale in telecommunications equipment development. This book focuses upon the international standard and deliberately avoids details of national variants. In this way, the principles of CCS systems can be explained in an independent manner.

This book is not intended as an exhaustive account of CCS systems: this is left for the specifications. So, whereas the specifications need to define a vast range of failure modes and error conditions, these are largely omitted from this description to enable the reader to concentrate on the principles. Once the principles have been absorbed, the specifications will be much easier to read. In the same way, examples are given throughout the book rather than an attempt being made to define comprehensively particular aspects of the signalling systems.

1.3 Channel-associated signalling

In CAS systems, the transfer of signalling information is conducted over signalling capacity that is specific to a particular circuit. Fig. 1.3 illustrates the concept of CAS systems. For each speech circuit from the customer to

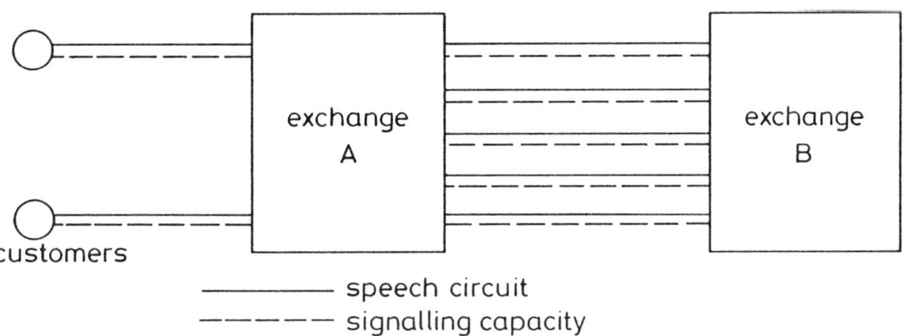

Fig. 1.3   Concept of channel-associated signalling

Exchange A, there is a directly-associated signalling channel. Numerous speech circuits exist between Exchanges A and B, and each speech circuit has a directly-associated means of transferring signalling information. There is a physical tie between the signalling path and the speech circuit. The signalling information can be carried in several forms, outlined in items (a) to (d) below.

(*a*) In loop-disconnect signalling systems, electrical conditions that apply to an analogue circuit can be varied to signify different meanings. For example, an electrical loop between a customer and a local exchange can be made and broken to signify the need to establish and release a call respectively. Trains of timed loop disconnects can indicate digits being dialled by the calling customer.

(*b*) In voice-frequency signalling systems, specific frequencies can be transmitted down analogue speech circuits in certain orders or in timed pulses to indicate a range of signals. These systems can use a single frequency or a combination of frequencies to define a signal.

(*c*) Outband-signalling systems can be used in analogue transmission links carrying multiplexed circuits. Such transmission links often allocate 4,000 Hz for each circuit but only use the range 300-3,400 Hz for speech. A typical outband-signalling system uses the frequency 3,825 Hz to define signals.

(*d*) In digital transmission links, the circuits are typically 64 kbit/s streams. One method of transferring CAS information on digital transmission links is to encode the frequencies that have been described above into digital format and transmit that information down the speech path to which the information refers. Another method is to allocate one of the 64 kbit/s streams as a signalling channel and allocate a part of that signalling channel to each speech circuit within the transmission link. In this case, there is still dedicated signalling capacity for each speech circuit, but it is collected together and placed into a single signalling channel.

One consequence of the direct association between signalling and speech paths in analogue CAS systems is the need to provide a signalling generator and a signalling terminator for each speech circuit at each exchange. This per-circuit provision of signalling equipment results in an inherent cost penalty. This disadvantage is further explained in Section 1.4.

CAS systems transfer three types of information:

(i)   Line signalling,
(ii)   Selection signalling and
(iii) Network signalling.

The purpose of line signalling is to convey information that changes the status of the speech circuit to which the signal refers. Basic line signalling covers the set-up and release of a speech circuit. Examples of line signals are:

Seize: indicating a request to use a particular speech circuit
Answer: indicating that the called customer has answered the call
Clear forward: indicating that the calling customer has released the call
Clear back: indicating that the called customer has ceased the call.

Selection signalling is used to convey address information (e.g. the digits dialled by the calling customer to identify the called customer). The address information is used by the network to route the call to the called customer. Depending on the signalling system used, selection signalling may also include procedures to optimise the process of transferring address information through the network. One measure of the quality of service perceived by the calling customer is the 'post-dialling delay'. The post-dialling delay is the time between completion of dialling the address information and the receipt of a tone (e.g. ring tone) by the calling customer. The techniques used within the network for selection signalling have a great impact upon the post-dialling delay. The need to minimise the post-dialling delay has been a major contributor to the evolutionary process of selection signalling.

Network signalling covers those aspects of information transfer relating to network operation, e.g. maintenance and operational support information. Whilst line signalling and selection signalling apply to both inter-exchange and access signalling systems, network signalling applies primarily to inter-exchange signalling systems.

Chapter 2 gives more information on the principles of CAS systems and describes six categories of CAS system. National variations are prolific, but the six categories summarise the basic characteristics of most systems used worldwide.

1.4 Common-channel signalling

1.4.1 Principles

In common-channel signalling (CCS) systems, the physical tie between the signalling path and the traffic circuit is removed. All signalling transfer relating to a transmission link takes place over a dedicated signalling channel (Fig. 1.4). Hence, a common-signalling channel handles the transfer of signalling information for numerous traffic circuits. Signalling capacity is not reserved for each traffic circuit, but signalling capacity is allocated dynamically as and when required. Fig. 1.4 shows the concept of CCS for both access and inter-exchange signalling. Exchanges A and B are connected by numerous speech circuits, denoted by solid lines. All the signalling that relates to the speech circuits is transferred between the exchanges using the common-signalling path (denoted by a dotted line). The common-signalling path can be regarded as a pipe between two exchanges, typically operating at 64 kbit/s, into which all signalling information is funnelled. Similarly, all signalling information pertaining to the speech circuits between each customer and Exchange A is transferred via the access signalling channel.

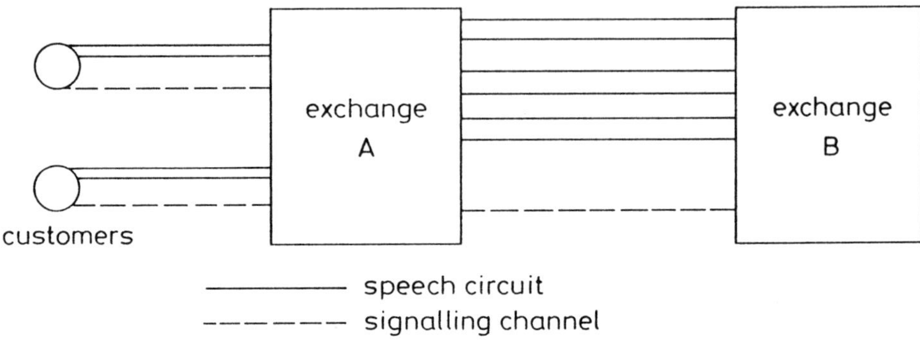

Fig. 1.4   Concept of common-channel signalling

The transfer of signalling information is achieved by sending 'messages' down the common-signalling path. A message is a block of information that is divided into fields, each field defining a certain parameter. The structure of a message, including the fields and parameters, is defined by the specification of the signalling system. The use of messages in CCS systems opens up a whole range of flexibility that is not present in CAS systems. Instead of being limited to a small number of meanings for signals, messages can be designed to cover a multitude of situations and services. This approach is a major advantage of CCS systems.

Numerous examples of messages are given in later chapters, but an example of a simple message is given in Fig. 1.5. In the example, Field 1 contains a 'flag', which is a unique code that identifies the start of the message. Hence, an exchange receiving numerous messages from the common-signalling channel can detect when one message finishes and another message starts.

Because there is not a dedicated relationship between traffic circuits and signalling, all signalling information relating to a transmission link being funnelled into one signalling channel, it is necessary to identify to which traffic circuit a particular message refers. This is achieved by including information within the message itself and this is the function of Field 2 (identifier) in Fig. 1.5. Originally, the main applications for which CCS systems were designed were to control the set-up and release of traffic circuits. For signalling purposes, each traffic circuit is allocated a unique number that identifies that traffic circuit. Information related to a particular circuit (e.g. an instruction to release a particular traffic circuit between two exchanges) is identified by the number of the circuit to which the information refers. This form of identification is known as 'circuit-related'. The identifier field in a circuit-related application is the number of the traffic circuit to which the message pertains.

field 4	field 3	field 2	field 1
check	information	identifier	flag

Fig. 1.5   Example of a simple message

One major advantage of CCS systems is that the signalling information is not restricted to the control of traffic circuits. The signalling channel can be used as a general data-transfer mechanism, e.g. to handle network administration information. Hence, in addition to the circuit-related application, CCS systems are required to transfer information that is not specific to a particular circuit. For this case, it is necessary to adopt a reference number approach to identify the relevance of a message. In this approach, a reference number is allocated by exchanges or customers' equipment from a pool of available numbers, thus identifying the message pertaining to a particular transaction. This form of identification is known as 'non-circuit-related' because there is not a pre-determined correlation between the identifier field and a particular circuit. The identifier field in a non-circuit-related application is a reference number allocated by exchanges or customers' equipment.

The information field in Fig. 1.5 contains the heart of the signalling information that is being transferred. For example, the information field could be coded 'release', meaning that the traffic circuit indicated in Field 2 should be disengaged. Field 4 contains check bits, which are generated by a known algorithm and which are used to ensure that the message is not corrupted during its passage through the network.

CCS systems make use of the intermittent nature of signalling for traffic circuits. Consider two customers who establish a basic telephone call. To establish the call requires signalling between the calling customer and the network, between exchanges within the network and between the network and the called customer. During the conversation phase of the call, there is no requirement to transfer signalling information. When the call is completed, there is again a need to transfer signalling information to release the speech circuit. This example illustrates the intermittent nature of signalling in basic telephony applications and introduces the concept of 'signalling activity'. The signalling activity when setting-up and releasing a circuit is high; however, on average the signalling activity for a circuit is low because there is no signalling when calls are not being made and during the conversation phase of a call. Hence, a single CCS channel can be used to handle numerous traffic circuits. The theoretical limit of the number of traffic circuits handled by a CCS channel is very high, but a typical practical value is 2000 traffic circuits. The picture becomes more complex when non-circuit-related signalling activity is taken into account. Non-circuit-related signalling can be intermittent (e.g. if it is used during call establishment to interrogate a database) or it can exhibit a high average signalling activity (e.g. if it is used to transfer large amounts of management data between nodes in the network). Hence, in a practical network, the calculation of signalling activity is an important element in designing the structure of the network and dimensioning signalling links.

In CAS systems, signalling capacity is dedicated to a traffic circuit. Limitations exist on when signals can be sent, depending on the status of the call. For example, it is not possible to send voice-frequency signals during the speech phase of a telephone call in some CAS systems, unless special measures are taken (e.g. provision of filters), because the customers would be able to hear the tones. However, within these constraints, it is possible to send signals instantaneously. In CCS systems, the philosophy is different: each message takes up the whole of the signalling channel for a very short length of time. Hence, it is not possible for an exchange to send two messages relating to two circuits from a transmission link at exactly the same time, albeit that each message might only take a few microseconds to transmit. For this reason, 'buffers' are provided at each end of a CCS link to store each message until the link becomes available. As messages are generated by an exchange, they are stored in the buffer and transmitted in a specified order. A typical order of transmission is first-in/first-out, resulting in messages being transmitted in the same order as they are received. When there are no messages to transmit, there is a need to maintain synchronisation of the signalling channel between two exchanges. This is achieved by continuously transferring synchronisation information until a new message is ready for transmission.

Because a small error in a message could change its meaning dramatically, and it is important in an environment of software-controlled nodes to ensure that logical sequences are followed accurately, error detection and correction mechanisms are employed for each message sent. It has already been stated that check bits are included in each CCS message: further details of detection/correction mechanisms are given in later chapters. To improve signalling reliability on a network-wide basis, facilities such as network reconfigurations are provided. For example, if a signalling link between two a transmission link over other transmission links and even via other exchanges. These facilities are carefully defined to ensure that messages are received by the intended destination in the correct order.

CCS systems are specified in terms of 'formats' and 'procedures'. The specification of the formats defines the structure of the messages used and the meaning of each field within the message. The specification of the procedures defines the logical sequence in which messages can be sent. The procedures of CCS circuit-related systems are very closely linked to the functions within exchanges that control the set-up and release of calls. There is, therefore, a close relationship between CCS procedures and exchange call control and a major element in defining CCS systems is the need to achieve an optimum balance between these factors.

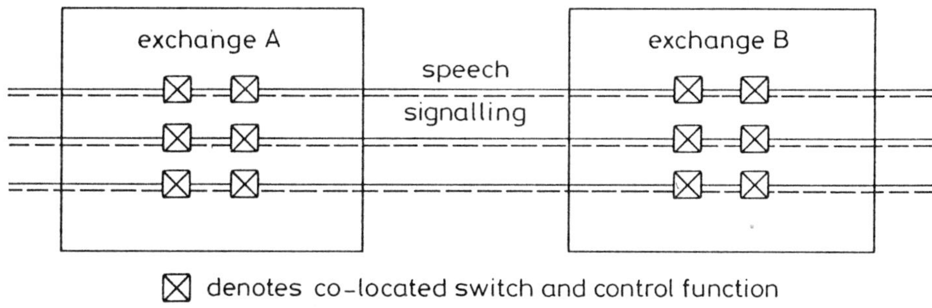

⊠ denotes co-located switch and control function

Fig. 1.6   Signalling in step-by-step exchanges

1.4.2 Evolution of signalling systems

One of the major factors influencing the development of signalling systems is the relationship between signalling and the control function of exchanges. Early telecommunications networks used analogue step-by-step exchanges. In such systems, the exchange constitutes a large number of discrete switches, each switch having a control mechanism vested in the switch itself. This means that, in the case of a telephone number dialled by a calling customer, discrete switches within a local exchange act upon individual digits to route the call: the control function, or decision-making function, is located in each discrete switch. In concept, this can be shown as the control and switch function being co-located, as illustrated in Fig. 1.6. In this type of exchange, when a call is made, the signalling and traffic follow the same path within the exchange. Step-by-step exchanges are invariably associated with

CAS systems; hence the signalling and traffic also follow the same path external to the exchange, i.e. on the transmission link.

The next stage through which exchanges evolved is shown in Fig. 1.7. In this case, the discrete switches of the step-by-step exchange are replaced by a 'switch-block' through which calls are routed. The control mechanism in the exchange for setting-up and releasing calls is separated from the switch block. This technique allows much more flexibility in controlling calls and it also reduces costs. Again, CAS systems are typically associated with this type of exchange. Whereas signalling information is carried on the same path as its associated speech circuit external to the exchange, the two are separated within the exchange. This is shown in Fig. 1.7, in which the speech traffic

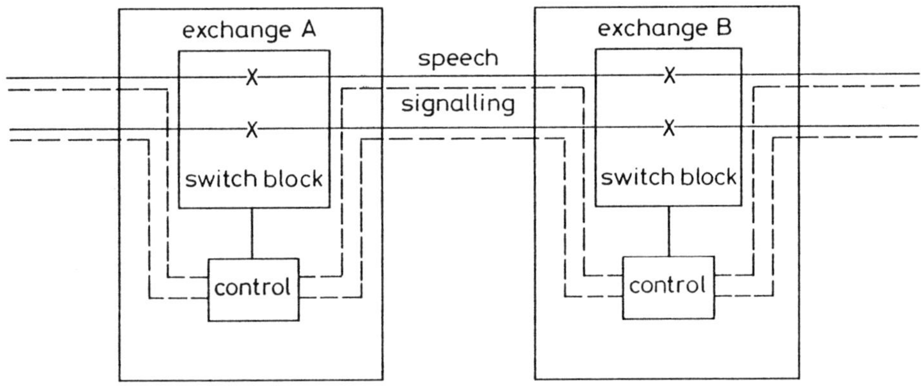

Fig. 1.7   CAS signalling with separate control and switch blocks

circuits (denoted by solid lines) are routed by the switch block but the signalling information (denoted by dotted lines) is routed via the control function. Between Exchanges A and B, the signalling and traffic are carried over the same path. This approach was primarily designed to allow optimisation of functions within exchanges, but its effectiveness is con-strained by the need to combine the signalling and speech traffic external to the exchange.

With CCS systems, the philosophy is to separate the signalling path from the speech path. This separation occurs both within the exchange and external to the exchange (Fig. 1.8), thus allowing optimisation of the control processes, switch block and signalling systems. Fig. 1.8 illustrates that, in a CCS environment, the speech paths are routed by the switch block, as before. However, the signalling (denoted by a dotted line), is routed by a separate path, both internal and external to the exchange. This approach allows maximum flexibility in optimising exchange and signalling development. The approach gains maximum benefit when adopted in parallel with the introduction of digital exchanges and digital transmission systems, CCS systems being particularly efficient in these circumstances. Whilst CCS systems can be used in an analogue environment, they are a major part in the drive towards all-digital networks. Networks offering end-to-end digital communications between customers are termed 'integrated services digital networks (ISDNs)'. It is in ISDNs that the full benefits of CCS systems can be unleashed.

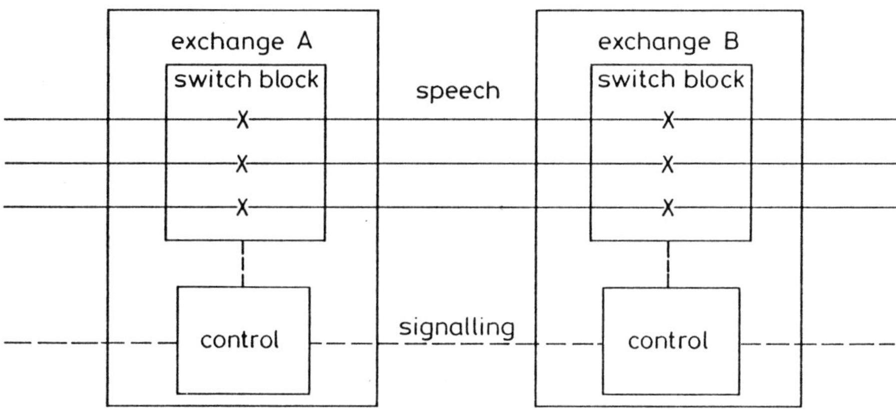

Fig. 1.8   Common-channel signalling

1.4.3 Advantages of common-channel signalling

CCS is being adopted throughout the world in national and international networks for numerous reasons. The reasons can be categorised into:

(*a*)   The rapidly changing control techniques of exchanges
(*b*)   The limitations of CAS systems
(*c*)   The evolutionary potential of CCS systems.

One result of the evolutionary process of exchanges described above is to change the relationship between signalling and call control. In the early exchange systems, exchanges could communicate, but in a limited and inflexible manner, thus limiting the flexibility of call control. In a CCS environment, the objective is to allow uninhibited communication between exchange control functions, or processors, thus tremendously broadening the scope and flexibility of information transfer. There is a close relationship between CCS and call control. In circuit-related applications, procedures for CCS systems define the logical sequences of events that occur during call establishment and release, thus having a direct impact upon the call control process. Thus, both CCS and call control have to be considered when an attempt is made to optimise one or the other. In some non-circuit-related applications (e.g. Transaction Capabilities described in Chapter 7), there is a desire to separate the call-control functions from the signalling system with the aim of reducing the dependence of each upon the other. Whilst this is successful to a degree, it is still necessary to consider call control and CCS in conjunction with each other to ensure that compatibility exists.

Further advantages result from the evolutionary process of CCS and call-control. The drive to provide an unrestricted communication capability between exchange processors eliminates per-circuit signalling termination costs. These costs are inevitable in per-circuit CAS systems, but by funnelling all signalling information into a single common-channel, only one signalling termination cost is incurred for each transmission link. There are cost penalties for CCS systems; e.g. the messages received by an exchange have to be analysed, resulting in a processing overhead. However, these cost penalties are more than covered by the advantages of increased scope of inter-processor communication and more efficient processor activity.

The separation of CCS from traffic circuits, and the direct inter-connection of exchange processors, are the early steps in establishing a cohesive CCS network to allow unimpeded signalling transfer between customers and nodes and between nodes in the network. The concept of a cohesive CCS network opens up the opportunity for the implementation of a wide range of network-management, administrative, operations and maintenance functions. A major example of such a function is the quasi-associated mode of operation (described in Section 1.6). This mode of operation provides a great deal of flexibility in network security, reduces the cost of CCS on small traffic routes and extends the data-transfer capabilities for non-circuit-related signalling.

CAS systems possess limited information-transfer capability due to:

(i)   The restricted number of conditions that can be applied (e.g. the limited variations that can be applied to a D.C. loop or the limited number of frequency combinations that can be implemented in a voice-frequency system) and

(ii) The limited number of opportunities to transfer signals (e.g. it is not possible to transmit voice-frequency signals during the conversation phase of a call without inconveniencing the customers or taking special measures).

Neither of these restrictions apply to CCS: the flexible message-based approach allows a vast range of information to be defined and the information can be sent during any stage of a call. Hence, the repertoire of CCS is far greater than channel-associated versions and messages can be transferred at any stage of a call without affecting the calling and called customers.

CCS systems transfer signals very quickly. A message used to establish a call in a CCS system can contain all the address digits in an information field. The message is delivered in a form suitable for modern processor-controlled exchanges, thus resulting in fast route selection. This speedy signalling also permits the inclusion of far more information without an increase in post-dialling delay.

Techniques used in modern CCS systems can further improve the flexibility provided to customers. 'User-to-user' signalling is a technique whereby messages can be transferred from one customer to another without undergoing a full analysis at each exchange in the network. Similarly, 'end-to-end' signalling allows exchanges to transfer information to each other without intermediate exchanges having to fully process the messages. Whilst forms of end-to-end signalling are possible using CAS systems, the technique can be more efficiently implemented with CCS systems. Further information on user-to-user and end-to-end signalling is given in Chapter 6.

One of the problems that prompted the development of CCS systems was 'speech clipping' in the international network. In some CAS systems, it is necessary to split the speech path during call set-up to avoid tones being heard by the calling customer. This results in a slow return of the answer signal and, if the called customer starts speaking immediately after answer, then the first part of the statement by the called customer is lost. As the first statement is usually the identity of the called customer, this causes a great deal of confusion and inconvenience. CCS systems avoid the problem by transferring the answer signal quickly.

As a result of the processing ability of CCS systems, a high degree of reliability can be designed into the signalling network. Error detection and correction techniques can be applied with a resulting high confidence in the transfer of uncorrupted information. In the case of an intermediate exchange failure, re-routeing can take place within the signalling network, enabling signalling transfer to be continued. Whilst these features introduce extra requirements (see Section 1.4.4), the common-channel approach to signalling allows a high degree of reliability to be implemented economically.

A major restriction of CAS is the lack of flexibility, e.g. the ability to add new features is limited. One factor that led to the development of CCS was the increasing need to add new features and respond to new network requirements. Responses to new requirements in CCS can be far more rapid and comprehensive than for channel-associated versions.

CCS systems are not just designed to meet current needs. They are designed to be as flexible as possible in meeting future requirements. One way of achieving this objective is to define modern CCS systems in a structured way, specifying the signalling system in a number of tiers. Chapter 4 describes the structure of modern CCS systems. This approach allows each tier to be optimised for various applications, whilst avoiding changes to other parts of the signalling system. The result is a flexible signalling system that can react quickly to evolving requirements. Some of these requirements were foreseen in original designs of CCS system, but many additional requirements will arise.

A corollary of the structured approach is that modern CCS systems are not restricted to specific services (e.g. the establishment and clear-down of telephone calls). The ability of CCS systems to transfer general data, and the increased range of messages that can be transferred, mean that information related to any service can be handled. Hence, future services can be incorporated in a flexible and comprehensive manner. Changes to existing services can be implemented more quickly and at lower cost than with CAS systems. A key element in the flexibility of CCS systems is the drive towards a signalling network providing unimpeded transfer of information and separate from speech traffic. The significance of this separation becomes even greater for non-circuit-related uses. This separate signalling network is the key to the future flexibility of telecommunications networks. This is the reason for the critical importance of CCITT Signalling Systems No 7 and DSS1 in telecommunications networks.

1.4.4 Requirements of common-channel signalling

The introduction of CCS has many advantages, but additional requirements are introduced in three areas:

(*a*) Reliability and security,
(*b*) Speech continuity and
(*c*) Processing overhead.

A signalling channel carried on a 64 kbit/s link has the practical capacity to control approximately 2000 traffic circuits. Hence, the failure of an inter-exchange signalling link would cause the loss of a significant amount of speech traffic. For access signalling, the loss of the signalling link would mean isolation of the customer from the local exchange. It is therefore essential to take exceptional precautions to avoid such losses. On a message basis, error detection and correction mechanisms have already been described and more details are given in Chapters 5 and 8.

Signalling security can also be improved by developing the signalling network itself. In the access network, it is possible to provide two signalling links to a customer (preferably on physically-diverse transmission links) and to switch all signalling traffic to one link when the other link is interrupted. Similar arrangements can be made for inter-exchange signalling, with automatic reconfiguration of signalling paths, even via different exchanges, to maintain a signalling continuity in the event of the failure of a signalling link.

Using these techniques, security in the signalling network can be enhanced to meet the CCITT requirements for the unavailability of signalling. For example, the unavailability of signalling communication between two exchanges is specified as a maximum of 10 minutes per year [2].

The message and network techniques for improving security of signalling information are positive assets. However, there is a cost in implementing such techniques and these must be taken into account when assessing the advantages of CCS.

CAS systems that use the speech path to transfer signalling information provide the inherent feature of checking the continuity of the speech path being established before conversation begins. If continuity is not achieved, the signalling transfer is not successful and the call is aborted or a further attempt is made to connect the call. This inherent continuity check is absent in CCS systems, owing to the separation of the speech and signalling paths. Hence, if considered desirable, separate speech-continuity checks can be provided. However, many modern digital exchange and transmission systems do not require such a continuity check because of the general level of equipment reliability, and inherent self-checking features, in such systems.

The flexible manner in which CCS systems are structured and the implementation of complex network-management features mean that extra processing is necessary to operate CCS. Even the inherent concept of funnelling all signalling on a transmission link into a common-signalling channel means that messages must be analysed to determine to which circuit (or transaction) they refer. However, this extra processing overhead is more than outweighed by the benefits of CCS systems.

1.5 Differences between access and inter-nodal CCS systems

The first two CCS systems to be defined internationally, CCITT Signalling Systems No 6 and 7, were inter-nodal systems for use within networks.

The objective of telecommunications networks is to provide fully-flexible high-capability communication between customers. To achieve this objective not only requires evolution of the network but also requires corresponding abilities in links to customers. Hence, a parallel transition is taking place for access signalling systems. Since the early inter-nodal CCS systems were specified, there has been a progressively wider application of CCITT specifications, recognising the advantage of economies of scale in terms of development and manufacture of signalling systems. This is particularly the case as the systems become more complex. Thus, CCITT now specifies an access CCS system for use between customers and the network. The system is called 'Digital Subscriber Signalling System No. 1 (DSS1)'.

Access CCS systems must be able to interwork effectively with network CCS systems. Hence, there are many common aspects between the two systems and the principles described so far are applicable to both types of system. However, some differences between the design of access and inter-nodal systems arise, as described below.

CCS systems continuously evolve to provide an increasing range of services, but nodes within a telecommunications network cannot be upgraded simultaneously. Hence, at any particular time, the degree of complexity of CCS system that one node can handle might not be the same as the degree of complexity that another node can handle. For example, some exchanges in a network might operate an early version of CCITT No. 7, whereas other exchanges might operate a later version. However, even though the degree of complexity at each node can vary, the general level of intelligence is similar (e.g. each node can provide a basic service, even though some nodes are less complex than others). This means that assumptions can be made about the basic capability of each node in the network when designing an inter-nodal CCS system. For access CCS systems, the capabilities of two communicating entities (e.g. a customer and a local exchange) can vary far more than for inter-nodal signalling systems. The exchange to which the customer is connected has at least the basic intelligence level discussed above. However, the capability of the terminal used by the customer can vary tremendously. The terminal can be a very simple telephone with a low level of intelligence or it can be a complex private automatic branch exchange (PABX) with as much intelligence as a network node. Furthermore, a network operator is often unaware of the type of terminal being used by a customer. Hence, access CCS systems must take account of this wide variation in intelligence of customer terminals, e.g. by allowing simple terminals to ignore complex information without affecting the ability to establish a basic telephone call.

There is a strong relationship between call-control and circuit-related applications of a CCS system, as explained further in Chapter 4. In inter-nodal CCS systems, a balance has to be achieved between the complexity of the call control at each node and the efficiency of the signalling links between the nodes. For example, in the user parts of CCITT No. 7 described in Chapter 6, many of the commonly-used messages include fields that do not explicitly state the name or length of the field. In these cases, the name and length of the field are derived from the type of message. This approach reduces the amount of information included in the message, thus allowing more messages to be carried on a signalling link. However, the approach can increase the level of processing required in an exchange because it is necessary to derive information from the message that is not provided explicitly.

In access CCS systems, there is less need to optimise the efficiency of the signalling link itself. This is because the signalling controls a limited number circuits and there is less to gain by increasing the efficiency of the link at the expense of call-control complexity.

When establishing a call between two customers, signalling information can have significance to:

(*a*) The link between the calling customer and the originating local exchange
(*b*) The link between the called customer and the destination exchange,
(*c*) The network only
(*d*) A combination of the elements of (a) to (c).

Establishing a basic call, or invoking supplementary services, usually involves all four types of information, but the largest amount of information usually applies to items (a) and (b). Hence, access CCS systems have to be capable of handling such large amounts of data on a per-call basis.

A CCS system for use in networks needs to exhibit extensive network-control mechanisms: these need to control both traffic circuits and signalling channels. Hence, CCITT No. 7 contains complex network-control techniques. Although access signalling systems require management functions to allow operation of the access link, the management techniques are generally simpler than those adopted in the network.

Local exchanges are not usually aware of the type of terminal connected by customers to access links. Indeed, numerous terminals can be connected to one access link. The protocols for the access link therefore have to take into account the prospect of multiple terminals being connected to one link.

1.6 Modes of operation

CCS systems can operate in a number of modes within telecommunications networks. To understand the modes of operation, it is first necessary to explain some appropriate terminology. An exchange in a telecommunications network that operates CCS is termed a 'signalling point'. Any two signalling points with the possibility of signalling communication are said to have a 'signalling relation'. The realisation of the signalling relation is by sending signalling messages between the two exchanges. The path taken by the signalling messages is determined by the mode of operation. Hence, the mode of operation determines how signalling messages are routed between signalling points. The mode of operation can be 'associated', 'non-associated' or 'quasi-associated'.

In the associated mode of operation, the signalling messages pertinent to a particular signalling relation are transferred over transmission links directly connecting the relevant signalling points. In Fig. 1.9, Exchanges A and B have a signalling relation and the signalling link directly connects the two exchanges. Hence, Fig. 1.9 represents the associated mode of signalling.

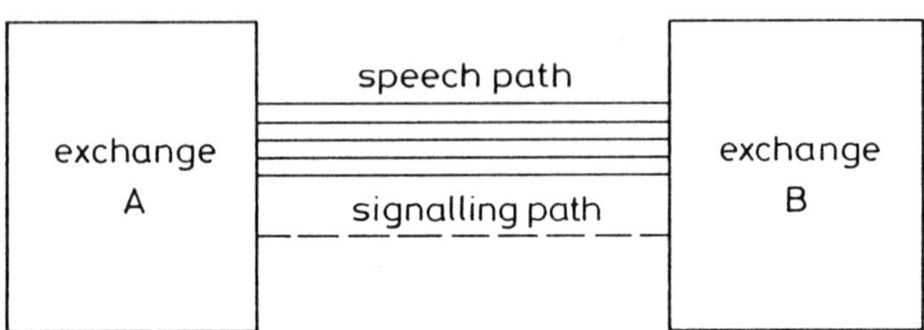

Fig. 1.9   Associated mode of operation

In the non-associated mode of signalling, the messages pertinent to a particular signalling relation are not transferred over transmission links directly connecting the relevant signalling points. Instead, the messages are transferred using intermediate (or tandem) signalling points. In an extreme case of non-associated signalling, each message between two exchanges could take a random route, with no preferred route being pre-determined by the network. However, in practical networks, a specific form of non-associated signalling is used, termed 'quasi-associated' signalling. In the quasi-associated mode of signalling, the path taken by a message through the signalling network is pre-determined by information assigned by the network.

In Fig. 1.10, Exchanges A and B have a signalling relation and are inter-connected by speech paths. However, the signalling path used to implement the signalling relation is via Exchange C (and not directly between Exchanges A and B). Hence, Fig. 1.10 is an example of the non-associated mode of operation. Because the message routeing through Exchange C is pre-determined by the network, Fig. 1.10 is also an example of the quasi-associated mode of operation. In this case, Exchange C is termed a 'a signal-transfer point (STP)'.

The quasi-associated mode of operation can be used as a back-up in the case of signalling link failure. For example, in Fig. 1.11, Exchanges A and B usually operate in the associated mode, with a signalling link directly connecting the two signalling points. However, in the case of failure of the A–B signalling link, the A–C–B signalling link can be used to control the speech paths between Exchanges A and B.

The quasi-associated mode of operation can also be used to reduce the overhead cost of CCS upon a small number of speech paths. In Fig. 1.12, Exchanges A and C and Exchanges C and B are connected by a large number of speech paths: Exchanges A and B are connected by a small

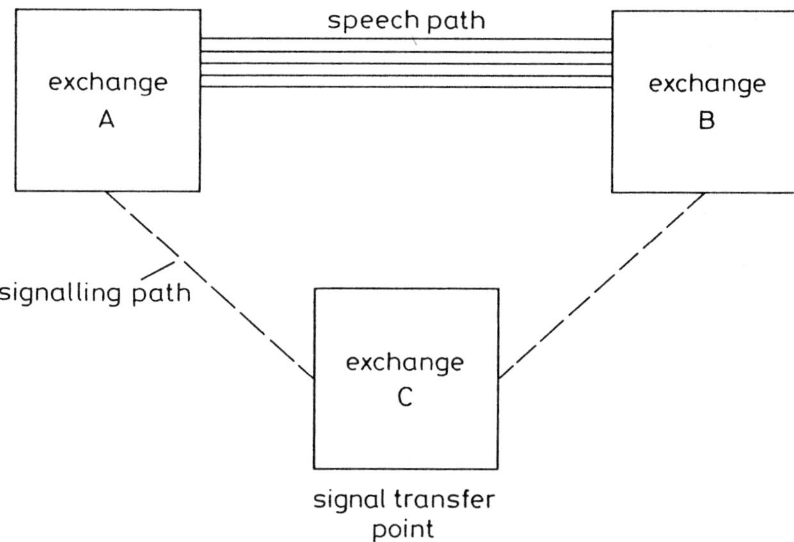

Fig. 1.10   Quasi-associated mode of operation

number of speech paths. In this case, the small number of speech paths between Exchanges A and B can be supported by quasi-associated signalling (routed A–C–B), thus avoiding the allocation of the signalling overhead to the A–B speech paths.

These modes of operation illustrate the great flexibility and powerful nature of CCS systems. Network features, like the quasi-associated mode of operation, greatly enhance the range of applications of CCS well beyond simple telephony call-control. Such features provide a major step towards CCS systems becoming general data-transfer mechanisms.

1.7 Chapter summary

Signalling is the vitalising influence of telecommunications networks. It transforms inert networks into powerful mechanisms for providing services to customers. There are two types of signalling system. In channel-associated signalling (CAS) systems, signalling capacity is provided on a dedicated basis for each traffic circuit. In common-channel (CCS) systems, signalling capacity is provided in a common pool and is allocated on a dynamic basis, as and when required.

Chapter 2 describes six categories of CAS system. However, modern networks are adopting CCS systems rapidly and this book concentrates primarily on describing CCS systems. International standards have been derived for the use of CCS systems in both national and international networks. These standards are described in this book to avoid reference to national variants. Chapter 3 gives a brief description of the CCITT Signalling System No. 6, but the focus of the book is on CCITT Signalling System No. 7 (CCITT No. 7) and the Digital Subscriber Signalling System No. 1 (DSS1). CCITT No. 7 is the modern CCS system for inter-nodal signalling and DSS1 is its counterpart for access signalling.

CCS systems use a common-signalling channel to carry signalling information for numerous traffic circuits. The signalling information is carried in messages. Each message is delineated by a flag and the transaction to which a message refers is defined by an identifier. In circuit-related applications, the identifier used is the circuit number of the appropriate traffic circuit. In non-circuit-related applications, the identifier used is a reference number that is independent of traffic circuits. Each message contains an information field defining the meaning of the message and check bits that are used to detect message corruption during transmission. Messages are stored in buffers to await the opportunity to be transmitted.

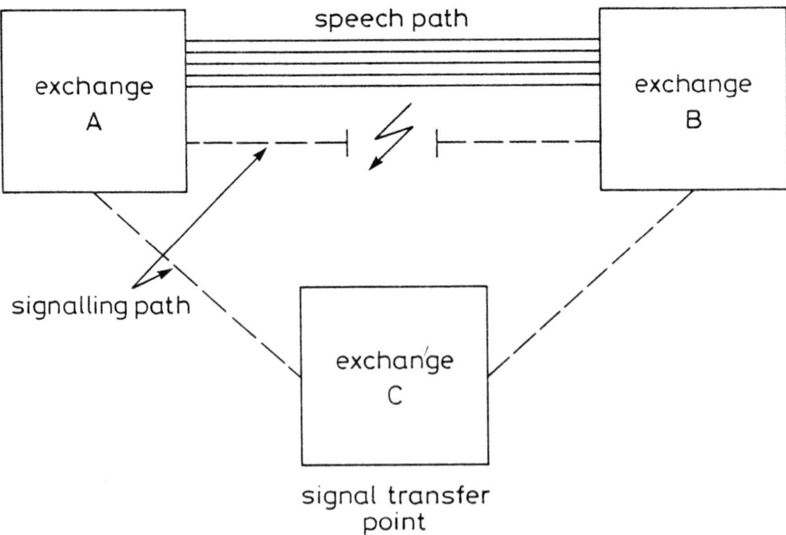

Fig. 1.11   Use of quasi-associated mode of operation for back-up

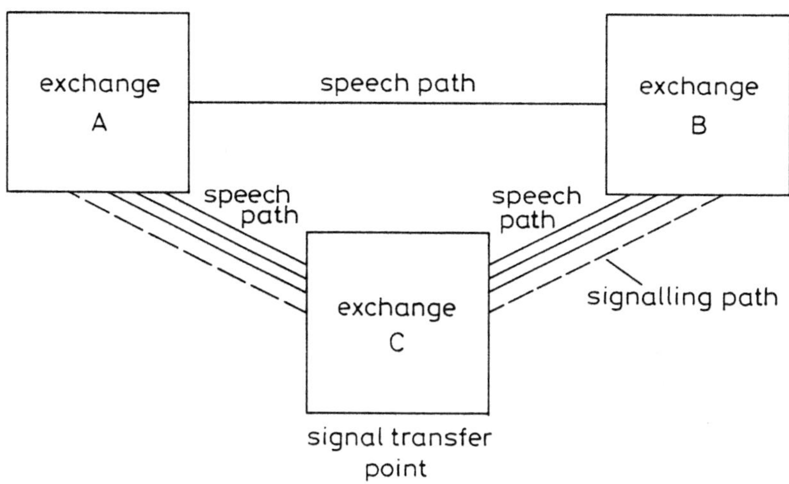

Fig. 1.12   Use of quasi-associated mode for small routes

The major advantages of CCS systems are that they are compatible with modern software-controlled networks, they overcome the limitations of CAS systems (particularly increasing the repertoire of messages that can be sent) and they exhibit evolutionary potential. Extra requirements are the need for a high level of reliability, the need to ensure traffic-circuit continuity and an increase in processing overhead.

Inter-nodal and access CCS systems have many aspects in common and such commonality is encouraged to reduce the complexity of interworking. However, the environments in which DSS1 and CCITT No. 7 perform are different. Thus, differences in characteristics arise, e.g. DSS1 must be able to handle a wide variety of customer terminals.

CCS systems can operate in a number of modes. The usual modes are the associated mode and the quasi-associated mode. In the associated mode, signalling messages are transferred over transmission links directly connecting relevant signalling points. In the quasi-associated mode, signalling messages are routed via signal-transfer points. The quasi-associated mode introduces a very flexible element into network design, e.g. quasi-associated signalling can be used as a back-up method of routeing signalling messages if disruption occurs to the normal signalling link.

1.8 References

1 WELCH, S: 'Signalling in telecommunications networks' (Peter Peregrinus, 1981).
2 CCITT Recommendation Q.706: 'Message transfer part signalling performance' (ITU, Geneva).

Chapter 2
Channel-associated signalling

2.1 Introduction

Chapter 1 describes the basic tenet of channel-associated signalling (CAS) systems; i.e. dedicated signalling capacity is provided for each speech circuit. In the past, standards organisations (e.g. CCITT) concentrated on the specification of international signalling systems, leaving national signalling systems to evolve in an independent manner. As new ideas were generated and new technology became available, national-network operators adopted new signalling systems. However, in the absence of international standards, each network operator optimised the various techniques for use in a particular network. Thus, whilst various categories of CAS system exist, the detailed application often varies from one network to another.

This chapter describes the principles of six categories of CAS system. Details are avoided, being well documented elsewhere [1], but examples are given to illustrate the techniques used.

At this stage, one point of terminology needs to be clarified. Signals can be sent in one of two directions. A 'forward signal' is sent in the direction from the calling customer to the called customer. A 'backward signal' is sent in the reverse direction.

2.2 Loop-disconnect signalling

In loop-disconnect signalling systems, the signalling information pertaining to an analogue speech circuit is transferred by modifying the electrical conditions applying to the speech circuit. The status of a direct-current loop on the speech circuit identifies the information being transferred. Loop-disconnect signalling is used to signal between customers and local exchanges and between exchanges in the network. For explanation, the link between the customer and the local exchange is described in this section.

An analogue speech circuit between a customer and a local exchange is based upon a direct-current loop. When a customer's telephone is in an idle state (i.e. the handset is in the cradle, 'on-hook'), the loop is disconnected (open) and current does not flow. When the calling customer commences a call (by lifting the handset, 'off-hook'), the loop is completed and a direct current flows. The current acts as a forward 'seize' signal at the originating local exchange, indicating that the customer wishes to make a call and reserving the speech circuit. The originating local exchange detects the loop from the calling customer and connects appropriate equipment to allow it to receive the dialled digits. As soon as the equipment is allocated by the local exchange, dial tone is returned to the calling customer. Dial tone is the invitation from the local exchange to the calling customer to start dialling. The dialled digits are communicated from the calling customer to the originating local exchange (i.e. in the forward direction) as a series of pulses by interrupting (breaking) the loop. Each digit is represented by a corresponding number of pulses. Hence, Digit 1 is represented by one pulse, Digit 2 by two pulses, etc. The normal rate of pulsing is at 10 pulses per second. Thus, to pulse the Digit 0 (10 pulses) takes one second, whereas to pulse Digit 1 takes one tenth of a second.

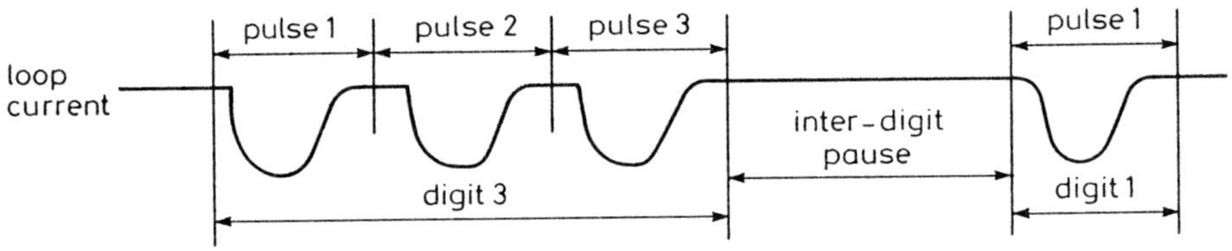

Fig 2.1   Principle of loop-disconnect signalling

Fig. 2.1 illustrates the principle involved. The format of each pulse varies in national networks, but a typical value is 66.7% break (i.e. open loop) and 33.3% make (i.e. closed loop). In Figure 2.1, there are three pulses in the first sequence. Thus, the local exchange recognises that Digit 3 has been dialled by the calling customer. There is an 'inter-digit pause' to allow the local exchange to recognise the end of one pulse sequence (i.e. digit) and the start of another. In this sequence, the inter-digit pause is followed by Digit 1. The dialled digits are analysed by the originating local exchange and the call is routed accordingly to the called customer at the destination local exchange.

When the called customer answers by lifting the handset, the loop between the destination local exchange and the called customer is closed. The answer state is transmitted through the network to the originating local exchange (i.e. the backward direction) by reversing the polarity of the loop within the network. When the calling customer ceases the call, by replacing the handset, the on-hook condition is communicated to the originating local exchange by means of a break in the loop. The break to indicate call completion is longer than the break associated with dialled digits. Hence, the originating local exchange can determine the difference between dialled digits and a request to clear the call. The request to clear the call is communicated through the network.

If the called customer indicates call completion by replacing the handset, the break condition is detected by the destination local exchange and signalled through the network by a reversal of loop polarity (i.e. return to the original loop polarity before answer). However, most telephony-based networks do not commence immediate release of a call in response to a request from the called customer: the calling customer is regarded as controlling the call. Thus, no action is taken by the network upon cessation of the call by the called customer unless fault conditions are detected (e.g. the expiration of a timer commenced upon the called customer ceasing the call). Examples of signals in a typical loop-disconnect signalling system are given in Table 2.1.

Table 2.1   Examples of signals in a loop-disconnect signalling system

Signal	Line Condition
Calling customer seizes	Loop to line
Dialled digits	Loop-disconnect pulses
Called customer answers	Reversal of loop polarity within network
Calling customer clears	Disconnection of loop

Loop-disconnect signalling combines the line and selection functions within one signalling mechanism. Signals from the calling customer are defined by make and break conditions on the loop. Signals from the called customer are transmitted through the network by monitoring line polarity. This type of signalling system is easy and cheap to implement, but the repertoire of signals that can be sent is very limited. The application of loop-disconnect signalling is also limited by the characteristics of the circuit upon which transmission is performed. A major restriction is the impact of line capacitance upon the pulse break. Line capacitance distorts the shape of the pulse, the distortion increasing with the length of the line. Equipment receiving the pulses can only countenance a limited degree of distortion before the reliability of recognising pulses is impaired. Thus, the line capacitance restricts the distance over which the signalling system can be used. This is exacerbated by the use of 'single-current working', in which the current varies between a positive value and a zero reference, as shown in Fig. 2.2. Because the waveform is not symmetrical about the zero reference, the pulse is susceptible to the distortion caused by line capacitance [1].

2.3 Long-distance DC signalling

Long-distance DC signalling systems attempt to overcome the distance limitations of loop-disconnect signalling systems by using 'double-current' working. Fig. 2.3 shows that, in double-current working, a pulse constitutes a reversal of current, the current varying between a positive and negative value. This approach of using a symmetrical waveform about a zero reference reduces the impact of pulse distortion due to line capacitance. In addition, double-current working reduces the degree of variance of pulse distortion in conjunction with varying signal level. Thus, the adoption of double-current working extends the distance over which the signalling system can be used.

Another advantage of double-current working is that sensitive receive devices can be implemented to detect the current. When using long

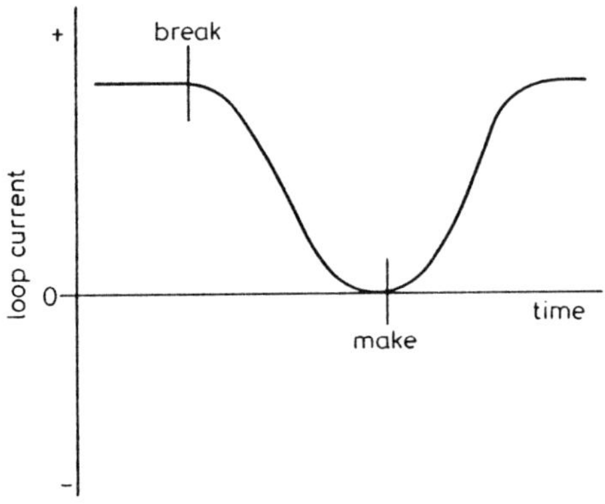

Fig. 2.2   Single-current working

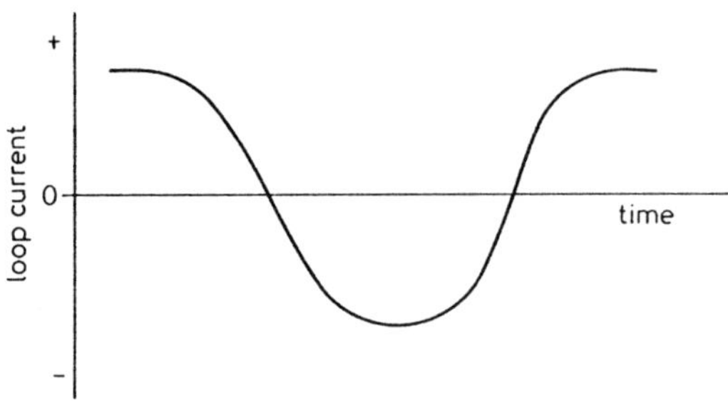

Fig. 2.3   Double-current working

signalling ranges, the associated current levels are low. Because of the nature of double-current working, it is easier to implement sensitive receive devices than it is to implement corresponding devices for single-current working.

The implementation of double-current working requires the adoption of features additional to those for single-current working. Thus, whilst long-distance DC signalling systems can be used over greater distances than standard loop-disconnect systems, long-distance DC systems are more expensive to implement. Examples of signals used in a typical long-distance DC signalling system [2] are given in Table 2.2.

Table 2.2 Examples of signals in a typical long-distance DC signalling system (UK DC2 system)

Signal	Line Condition
Idle	Negative battery on one leg, earth on the other leg.
Calling customer seizes	Loop current reversed
Dialled digits	Loop current reversals at 10 pulses per second
Answer	Earthed loop
Calling customer clears	Reversal of seize condition, i.e. return to idle loop

2.4 Voice-frequency signalling

Voice-frequency-signalling systems transfer signalling information over the speech circuit to which the information refers. Signal transfer is achieved by generating one or more tones and transmitting them over the appropriate speech circuit. A tone receiver at the other end of the circuit is used to analyse the information content. The signals share the same 300—3400 Hz frequency band as that used for the speech circuit. This means that signals cannot be sent during the conversation phase of a call unless, for example, expensive filters are provided to prevent the signal being heard by the calling and called customers. During the set-up phase of the call, the speech circuit is split to avoid the calling customer hearing the signals.

Between exchanges, the signals are treated as normal speech. Thus, speech amplifiers are used to maintain signal strength, resulting in a much greater signalling range than is possible in loop-disconnect and long-distance DC signalling systems. Tone receivers are permanently allocated to each circuit. Thus, the cost of implementation is relatively high. Voice-frequency systems can handle both line and selection signalling, but selection signalling is sometimes covered by other means, e.g. the multi-frequency-signalling system described in Section 2.6.

Several forms of voice-frequency signalling exist. They can be categorised into 'continuous-compelled', 'continuous non-compelled' and 'pulsed'.

In continuous-compelled systems, the information is transferred by sending a continuous frequency (or combination of two frequencies) during the call sequence. The term 'continuous' is used to denote that the frequency is transmitted without interruption, i.e. the frequency is not pulsed. The systems are known as 'compelled' because the application of the frequency is maintained until an acknowledgement is received. The signal meaning is determined by the direction of the signal, the frequency of the signal and the point during the call sequence at which the signal is sent. To illustrate the principles, examples of line signals in the CCITT Signalling System No. 5 [3] are given in Table 2.3.

Table 2.3   Examples of line signals in CCITT Signalling System No. 5

Signal	Direction	Frequency
Seize	Forward	1
Proceed to send (seize acknowledge)	Backward	2
Answer	Backward	1
Answer acknowledge	Forward	1
Clear forward	Forward	1 + 2 (Compound)
Release guard (acknowledgement of clear forward)	Backward	1 + 2 (Compound)

Frequency 1 = 2400 Hz
Frequency 2 = 2600 Hz

In continuous non-compelled signalling systems, the information being transferred is denoted in the change of state of the tone. The term 'non-compelled' means that an acknowledgment of a signal is not required before the sending of the signal is stopped. This form of signalling system [4] conveniently covers the 2-state on-hook/off-hook conditions by means of tone-on/tone-off. A convenient method of implementation is tone-on to indicate the idle condition. With this method, the action of off-hook automatically invokes the tone-off condition, thus avoiding the need for complex measures to avoid the signal being heard by the calling customer. Hence, an advantage of this approach is the ability to use simple signalling terminals. Examples of typical signals using this approach are given in Table 2.4.

Table 2.4 Examples of signals in continuous non-compelled signalling systems (Bell SF system)

Signal	Forward tone	Backward tone
Idle	On	On
Seizure	Off	On
Answer	Off	Off
Clear forward	On	N/A

In pulsed voice-frequency signalling systems, the information is transferred by timed pulses of tone. Hence, the meaning of the signal is determined by the direction of the signal, the length of the pulse and the point during the call sequence that the signal is sent. The advantages of the pulsed form of voice-frequency signalling are that a larger repertoire of signals is available (thus allowing more features), higher signalling levels are possible (due to the non-continuous nature of the signals) and there is less interference with the signals (again due to their non-continuous nature). However, the need to meet the tolerances required for effective signal recognition mean that the signalling terminals are relatively complex and hence expensive. Examples of signals in a typical pulsed voice-frequency-signalling system [5] are given in Table 2.5.

Table 2.5 Examples of signals in a typical pulsed voice-frequency-signalling system (UK AC9 System)

Signal	Tone pulse in ms (frequency 2280 Hz)
Seize	70
Digits	60
Answer	250
Clear forward	Greater than 700

Voice-frequency-signalling systems can be used in two modes: 'end-to-end' and 'link-by-link'. In the end-to-end mode, signals of relevance only to originating local exchange and destination local exchange in a connection can be transferred directly over the speech path, without converting to direct-current conditions at intermediate exchanges. Hence, in end-to-end mode, signals (e.g. the answer signal) can be transferred very quickly. In the link-by-link mode, each signal is converted to direct-current conditions at each exchange (including intermediate exchanges), thus slowing down the transfer of the signals. Link-by-link signalling is the most commonly-used mode because most networks have adopted a range of different signalling systems, thus reducing the likelihood of achieving an end-to-end mode of operation. Using link-by-link signalling, an intermediate exchange can receive a call request using a voice-frequency signalling system and transfer the call request to the next exchange using a different type of signalling system.

The use of the same physical path for speech and signals necessitates the avoidance of imitation of the signals by normal speech. If specific measures are not taken, the signal receivers could confuse normal speech with a signal. Several factors are taken into account to avoid imitation:

(*a*)   the signal frequency is chosen to avoid high energy levels in normal speech. This factor, in conjunction with other factors, such as filter characteristics, results in frequencies of 2000 to 3000 Hz being suitable;

(*b*)   a guard circuit is employed such that, if other frequencies are present at the same time as the signal frequency, the guard circuit can over-rule the signal recognition circuit;

(*c*)   some signalling systems (especially international systems) use tones of dual frequency rather than single frequency, thus reducing the chances of signal imitation.

2.5 Outband signalling

The term 'outband signalling' refers to signalling systems that use frequencies above the normal telephone speech path spectrum. Outband signalling is typically used in frequency-division-multiplex (FDM) transmission systems. These are systems in which numerous traffic circuits are assembled into blocks for transmission purposes by translating them into different parts of the frequency spectrum. The allocated frequency band per circuit is typically 4 kHz per circuit. The speech and signalling are transmitted in separate sub-bands, with speech being carried in the 300—3400 Hz band and signalling in the 3400—4000 Hz band. This allows speech and signalling to be transferred simultaneously and independently. The speech and signalling are separated at each exchange by means of appropriate filters. The frequency recommended by CCITT [6] for outband signalling is 3825 Hz. Outband signalling is converted to direct-current conditions at each exchange. Hence, it is link-by-link in nature.

Outband signalling overcomes a number of the inherent disadvantages of voice-frequency signalling. Advantages include the ability to signal during speech and the avoidance of measures to overcome the imitation of signals by normal speech. A disadvantage of outband signalling is that it can only be applied on transmission systems that permit a wider frequency spectrum than normal non-multiplexed transmission systems. Hence, it is usually limited to appropriate frequency-division-multiplex transmission systems.

The nature of outband signalling allows it to be used in numerous modes, including the continuous mode and pulsed mode described for voice-frequency signalling in Section 2.4. A common implementation is the continuous non-compelled mode, either using tone-on idle or tone-off idle.

The influences over whether to use tone-on idle or tone-off idle include transmission system overload considerations and the impact of transmission interruptions. These characteristics vary amongst the numerous types of network, resulting in the use of both forms of signalling. Examples of signals in a typical tone-off idle implementation are given in Table 2.6. It is shown in Table 2.6 that allowance has been made for selection signalling as well as line signalling. Whereas some implementations include selection-signalling, other implementations adopt separate selection-signalling arrangements. An example of a tone-on idle system is the line signalling for CCITT R2 (analogue) [7]. Examples of the signals defined for R2 are given in Table 2.7.

Table 2.6 Examples of signals in typical tone-off idle system (UK AC8 System)

Signal	Forward tone	Backward tone
Idle	Off	Off
Seizure	On	Off
Digits	Off	Off
	(Pulse breaks)	
Answer	On	On
Clear forward	Off	–

Table 2.7 Examples of signals in CCITT R2 line signalling

Signal	Forward tone	Backward tone
Idle	On	On
Seizure	Off	On
Answer	Off	Off
Clear forward	On	–

2.6 Multi-frequency inter-register signalling

Developments in the common control of exchanges, and the need to reduce post-dialling delays in networks, led to the implementation of multi-frequency inter-register signalling systems. The term 'multi-frequency' reflects the use of a variety of frequency tones to denote different signals. Registers are used in common-control exchanges to store and analyse routeing data. They are provided on a common basis, i.e. a single register provides the routeing data for numerous speech circuits. In this way, a register is used by a call only for the time to set up that call. The register is then made available to help set up other calls. The term 'inter-register' refers to the signalling between registers in two different exchanges. One concept is to separate the line-signalling functions from the selection and network-signalling functions. This allows a simple line signalling arrangement to be maintained, whilst allowing more complex selection and network functions to be provided. If these functions were provided on a per-circuit basis, then the cost would be prohibitive. However, by providing registers on a common basis, the functions can be introduced without unduly increasing the cost per circuit. Hence, multi-frequency inter-register signalling covers only selection and network functions. Line functions are covered by an appropriate line-signalling system.

The basic concept in multi-frequency inter-register signalling is that each address digit (or signal) can be coded by means of two frequencies (compounded) taken from a range of N frequencies. Hence, each address digit is represented by a code that is composed of 2 out of N frequencies. For $N=5$, up to 10 combinations can be coded and this is sufficient to define Digits 0—9. However, $N=6$ gives 15 coding possibilities and thus increases the signal repertoire to a more satisfactory level. Hence, the usual implementation in networks is $N=6$.

The advantages of inter-register multi-frequency signalling include:-

(*a*)   a fast transfer of address digits through the network, thus reducing post-dialling delay.

(*b*)   a wide repertoire of signals compared with other CAS systems, the repertoire being available for use in the forward and backward directions.

(*c*)   resilience to distortion, due to signal content being determined by frequency recognition;

(*d*)   an ability to be used in end-to-end mode or link-by-link (as described in Section 2.4), depending on the particular requirements for a specific network.

(*e*)   an ability to accommodate pulsed, compelled or non-compelled approaches, as described in Section 2.4.

Multi-frequency inter-register signalling is very flexible compared with other CAS systems and many variations have been adopted throughout the world, depending on the network characteristics. To illustrate the principles involved, consider the formatting of CCITT Signalling system R2 [7]. System R2 is a 2-out-of-6 frequencies, compelled, end-to-end signalling system exhibiting a range of forward and backward signals.

Each signal is formed by the compound of 2-out-of-6 frequencies, as shown in Table 2.8. Each frequency combination can have two meanings in both the forward and backward directions, as shown in Tables 2.9 and 2.10. This increases the signal repertoire of the system. In the forward direction, each frequency combination defines a signal in a Group I category and a Group II category. The Group I category is used until a request to change to Group II is received. Backward signals are categorised into Group A and

Group B. As for forward signals, each frequency combination can have two meanings; Group A meanings apply until Signal A3 or Signal A5, as defined in Table 2.10, is returned.

Table 2.8 Signal frequencies in System R2

Signal number	Forward signal frequencies (Hz)	Backward signal frequencies (Hz)
1	1380 + 1500	1140 + 1020
2	1380 + 1620	1140 + 900
3	1500 + 1620	1020 + 900
4	1380 + 1740	1140 + 780
5	1500 + 1740	1020 + 780
6	1620 + 1740	900 + 780
7	1380 + 1860	1140 + 660
8	1500 + 1860	1020 + 660
9	1620 + 1860	900 + 660
10	1740 + 1860	780 + 660
11	1380 + 1980	1140 + 540
12	1500 + 1980	1020 + 540
13	1620 + 1980	900 + 540
14	1740 + 1980	780 + 540
15	1860 + 1980	660 + 540

Table 2.9 Examples of forward signals in System R2 Groups I and II

Signal number	Group I forward signal*	Group II forward signal
1	Digit 1	National use
2	Digit 2	National use
8	Digit 8	Data transmission
9	Digit 9	Customer with priority
15	End of pulsing	National use

* When a Group I signal is the first signal during call set-up on an international circuit, the meaning of the signal is different [7].

A typical sequence is for the originating exchange to send the first address digit (using Signals I–1 to I–10) to the destination exchange. This first digit is acknowledged by Signal A–1, which also requests the transfer of the next digit. After the first address digit, the remaining address digits are requested by the destination exchange using Signal A–1. When the last address digit had been received by the destination exchange, it returns a Signal A–3, indicating the need to changeover to Group B and Group II signals. If the called subscriber is free, a B–6 Signal is returned to the originating exchange.

Table 2.10   Examples of backward signals in System R2 Groups A and B

Signal	Group A backward signal	Group B backward signal
1	Send next digit	National use
3	Address complete, changeover to use of Group B and Group II signals	Customer busy
4	Congestion	Congestion
5	Send calling customer's category	Vacant national number
6	Address complete, set up speech conditions	Customer line free

2.7 Signalling in pulse-code-modulation systems

2.7.1 General

Pulse-code modulation (PCM) is a method of converting information from an analogue form to a digital form for transfer over digital transmission systems. The technique [8] involves sampling the analogue waveform and coding the results in a digital format. Successive sampling allows the analogue waveform to be represented by a series of 8-bit codes. The 8-bit codes from numerous speech channels are assembled into blocks for transmission by insertion into 'time-slots'. The technique is called time-division multiplexing (TDM). The structure of the time slots depends upon the standard adopted: the two usual standards are described below.

The term 'pulse-code-modulation (PCM) signalling' is in common use, but this is misleading and care must be taken to discriminate between the provision of a transmission medium and provision of corresponding signalling. A PCM system merely provides a digital transmission medium in which signalling capacity is provided. The signalling systems described earlier are carried in the signalling capacity made available by the PCM system.

The bandwidth required to transmit signals is much less than that for speech, so the signalling for several speech channels in a PCM system can be handled by a small portion of the bandwidth. The signalling capacity can be used for CAS or CCS. For CAS, the means of identifying to which speech channel a particular signal refers is to divide the signalling capacity into dedicated bit locations. Signals pertinent to a particular speech path are always transmitted in signalling-bit locations dedicated to that speech channel. The means of conveying CCS is to compound the signalling capacity into a signalling channel that is available as and when required.

The CCITT has defined PCM standards for 30-channel and 24-channel systems. The capacity available for signalling in these two standards is different as a result of the differing constraints applied by the PCM standards.

2.7.2 30-Channel PCM Systems

In 30-channel PCM systems, the 8-bit codes relating to 30 speech channels are time-division multiplexed into a 'frame'. Each 8-bit code is inserted into a time-slot within the frame, as shown in Fig. 2.4. Time-slot 0 is used for alignment purposes and Time-slots 1—15 and 17—31 are used for the encoded speech relating to the 30 channels. Time-slot 16 is dedicated for the use of signalling. Sixteen frames (Frames 0—15) constitute a 'multi-frame'.

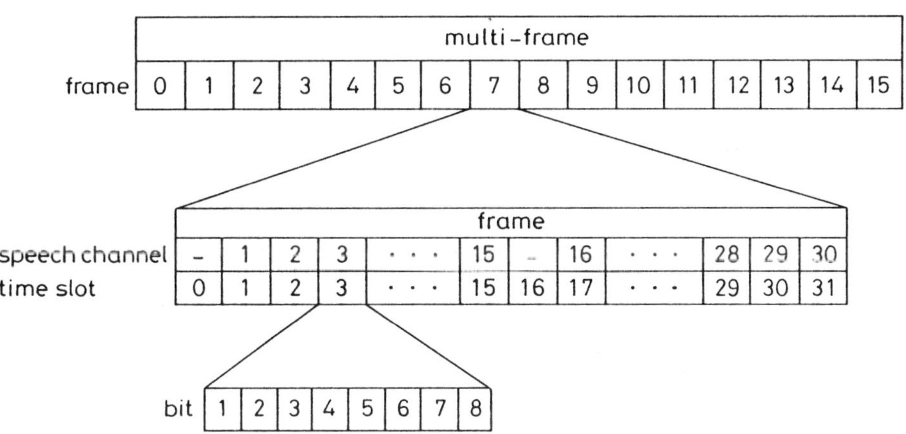

Fig. 2.4 Frame structure of 30-channel PCM system

The tenet of CAS systems is that dedicated signalling capacity is available for each speech circuit. This is achieved in 30-channel PCM systems by allocating 4 bits in each multi-frame to the signalling for each speech channel. This is illustrated in Fig. 2.5. Time-slot 16 in Frame 1 contains the signalling bits for Speech Channels 1 and 16. Time-slot 16 in Frame 2 contains the signalling bits for Speech Channels 2 and 17. This process is repeated to the end of the multi-frame, in which the Time-slot 16 in Frame 15 contains the signalling bits for Speech Channels 15 and 30. The next multi-frame repeats this pattern, starting with the signalling bits for Speech Channels 1 and 16 again. The coding of the *abcd* bits varies according to the implementation, but typical values [1] are given in Table 2.11.

For CCS systems, the signalling capacity in Time-slot 16 of successive frames is amalgamated to form a signalling channel. The multi-frame arrangements are not relevant for CCS systems. The capacity within the signalling channel is allocated dynamically according to need. For example, if Speech Channel 1 requires signalling information to be transferred, then the full capacity of the signalling channel is used to transfer the information. When completed, the capacity of the signalling channel is then used for another speech channel.

time–slot 16 of frame 1			time–slot 16 of frame 2			time–slot 16 of frame 15	
a b c d for channel 1	a b c d for channel 16		a b c d for channel 2	a b c d for channel 17		a b c d for channel 15	a b c d for channel 30

Fig. 2.5   Allocation of signalling capacity in time-slot 16.
abcd denotes the group of four bits allocated for the signalling of one speech channel

Table 2.11   Typical values of coding for signalling in Time-slot 16 (UK)

Forward signal	Code (*abcd*)	Backward signal	Code (*abcd*)
Idle	1111	Free	0111
Seize	0011	Answer	0011
Dialling	1011	Called customer clears	1111
Calling customer clears	1111		

2.7.3 24-Channel PCM Systems

Fig. 2.6 shows the frame and multi-frame arrangement for the 24-channel

PCM standard. In this case, each 8-bit code is again inserted into a time-slot, but there are only 24 time-slots in a frame. A multi-frame comprises 12 frames. Two blocks of signalling capacity are derived by designating Frames 6 and 12 as signalling frames. Bit 8 of the 8 bits in each time slot in Frames 6 and 12 is 'stolen' for signalling purposes. Hence, Frames 1—5 and 7—11 allow the full 8 bits for speech encoding, whereas Frames 6 and 12 allow 7 bits for speech encoding and 1 bit for signalling. This arrangement is illustrated in Table 2.12.

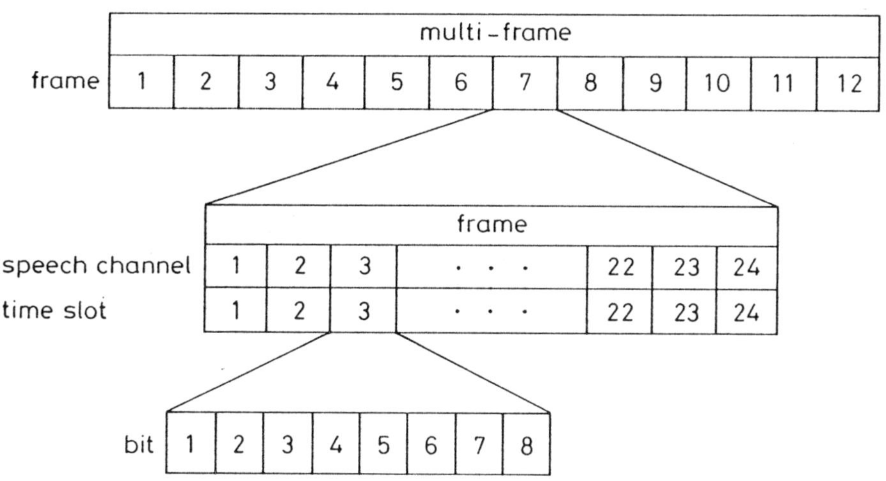

Fig. 2.6   Frame structure of 24-channel PCM system

This bit-stealing technique has a minimal effect upon the speech quality of the 24 speech channels, but it is a major constraint upon the ability to supply signalling capacity. The use of the stolen bits to specify line signals is illustrated in Table 2.13.

For CCS systems, signalling capacity can be provided in two ways. The first realisation is to use the first bit of successive even frames to convey the signalling information. As for the 30- channel systems, the multi-frame arrangements are not relevant and the resultant signalling capacity is amalgamated into a signalling channel. The second realisation allows a greater signalling capacity to be used by dedicating one of the 24 channels in the PCM system as a signalling channel.

Table 2.12   Signalling in the 24-channel PCM system (Bell D2)

Frame number	Bit number within each time slot used for:	
	Speech	Signalling
1	1–8	–
2	1–8	–
3	1–8	–
4	1–8	–
5	1–8	–
6	1–7	8
7	1–8	–
8	1–8	–
9	1–8	–
10	1–8	–
11	1–8	–
12	1–7	8

Table 2.13   Examples of coding for signalling channels in 24-channel PCM systems (Bell D2)

Forward signal	Value of signalling bit transmitted	Backward signal	Value of signalling bit transmitted
Idle	0	Idle	0
Seize	1	Answer	1
Calling customer clears	0	Called customer clears	0

2.8 Chapter summary

Lack of international standards for CAS systems in national networks has resulted in a large variety of implementations. However, CAS systems can be divided into six categories.

In loop-disconnect-signalling systems, signalling pertaining to an analogue speech circuit is performed by modifying the status of a direct-current loop on the speech circuit. The off-hook condition is denoted by closing the loop and the on-hook condition by breaking the loop. Dialled digits are communicated by timed interruptions to the loop. At the destination exchange, a reversal of loop polarity indicates that the call has been answered. Loop-disconnect signalling is effective for simple information, but the signal repertoire is limited and line capacitance distorts the pulses, thus limiting the signalling range.

Long-distance DC systems attempt to overcome the distance limitations of loop-disconnect systems by adopting double-current working. In this approach, a pulse constitutes a reversal of current, the current thus varying between a positive and negative value. The symmetrical waveform about a zero reference reduces pulse distortion due to line capacitance and allows sensitive receive devices to be implemented.

Voice-frequency-signalling systems transfer signalling information, over the speech circuit to which the information refers, by generating and receiving tones. In continuous-compelled systems, tones are continuously transmitted until acknowledged. The signal meaning is derived from the direction of transmission of the tone, the frequency of the tone and the point during the call sequence that the tone is transmitted. In continuous non-compelled systems, the signal meaning is derived from the state (on/off) of the tone. In pulsed systems, the information being transferred is included in timed pulses of tone and the signal is derived from the length of the pulse.

Outband-signalling systems are used in frequency-division-multiplex-transmission systems. In such systems, each speech circuit is typically allocated a frequency spectrum of 4 kHz, but only 300—3400 Hz is used for speech. Outband systems use a tone within the remaining frequency spectrum, typically 3825 Hz, to transfer signals. Outband signalling can be used in numerous modes, including continuous and pulsed.

Multi-frequency inter-register systems are designed to reduce post-dialling delays and increase the repertoire of available signals. They use two compounded tones, taken from a range of frequencies, to denote each address digit.

CAS systems can be carried within PCM transmission systems. In the 30-channel PCM system, Time-slot 16 is dedicated for signalling purposes. Four bits are provided in each multi-frame for the signalling pertaining to each speech channel.

In the 24-channel system, a bit-stealing technique is adopted. In this approach, 2 frames within a multi-frame are designated as signalling frames. In these frames, 7 bits are used for speech encoding (rather than the normal 8 bits) and the 8th bit is used for signalling. The stealing of bits for signalling has a minimal impact upon the quality of the speech in the traffic channels.

2.9 References

1 WELCH, S: 'Signalling in Telecommunications Networks' (Peter Peregrinus, 1979)
2 WELCH, S and HORSFIELD, B R: 'The Single Commutation Direct Current Signalling and Pulsing System', *Post Office Electrical Engineering Journal, 1951*, **44**, Part 1
3 CCITT Recommendations Q.140—Q.164: 'Specification of Signalling System No. 5' (ITU, Geneva)
4 BREEN, C and DAHLBOM, C A: 'Signalling Systems for Control of Telephone Switching', *Bell System Technical Journal*, 1960, **39**, pp. 1381–1444.

5 MILES, J V and KELSON, D: 'Signalling System AC No. 9', *Post Office Electrical Engineering Journal*, 1962, **55**, Pt. 1, 51–58.
6 CCITT Recommendation Q.21: 'General Recommendation on Telephone Switching and Signalling Functions' (ITU, Geneva)
7 CCITT Recommendations Q.350-Q.368: 'Specification of Signalling System R2' (ITU, Geneva)
8 CCITT Recommendations G.732 and G.733: 'Digital Networks Transmission Systems and Multiplexing Equipments' (ITU, Geneva)

Chapter 3

CCITT Signalling System No. 6

3.1 Introduction

CCITT Signalling System No. 6 [1] was the first CCS system to be implemented internationally. It was originally designed for use in the international network, but some flexibility is included to allow its use in national networks [2]. It is an inter-exchange signalling system. CCITT No. 6 has been implemented widely, but it is now being superseded by CCITT No. 7. Thus, a brief description of CCITT No. 6 is given in this chapter. Details of CCITT No. 7 are given in subsequent chapters. CCITT No. 6 offers a wide range of features associated with being a CCS system, including operation in the quasi-associated mode, error detection and correction mechanisms and re-routeing capabilities in fault conditions. However, its main drawback is its limited evolutionary potential. In a dynamic environment, CCITT No. 7 offers far greater flexibility and evolutionary capability, particularly due to its structured architecture (see Chapter 4).

CCITT No. 6 was originally optimised to be used over analogue transmission paths , but digital paths can also be accommodated. It is operated on a link-by-link basis, an end-to-end capability not being provided. Communication between two exchanges using CCITT No. 6 is via a data stream over a CCS link. The information sent over the link is divided into 'signal units' of 28 bits in length. All signal units are of the same length, the last 8 bits being used for error detection. The signal units are grouped into 'blocks' of 12, the last signal unit in each block being used for acknowledgement purposes. Each block is allocated a sequence number as part of the error detection and correction mechanism. Signal units within a block are either 'message-signal units' or 'synchronisation-signal units'. Message signal units contain, for example, call-control information relating to the control of telephone calls. Synchronisation-signal units are sent in the absence of signalling traffic to maintain synchronisation of the link.

Within an exchange, the functions performed in sending and receiving CCITT No. 6 signal units are illustrated in Fig. 3.1. The processor is the functional block responsible for call-control within the exchange. The processor therefore recognises the need to establish CCITT No. 6 signalling connections and conduct signalling communication. The processor formats appropriate messages and passes them to the output buffer. The output buffer stores each message until an appropriate slot is available to transmit the message to the next exchange. When it is the turn for a particular message to be transmitted, it is passed to the coder, which adds 8 check bits to the message. The check bits are used by the receiving exchange to detect corruption of the message. Further information on the use of the check bits is given in Section 3.5.1. In the analogue version of the signalling system, the signal is prepared for analogue transmission by a modulator and delivered to the appropriate outgoing signalling channel. In the digital version, modulation is not necessary.

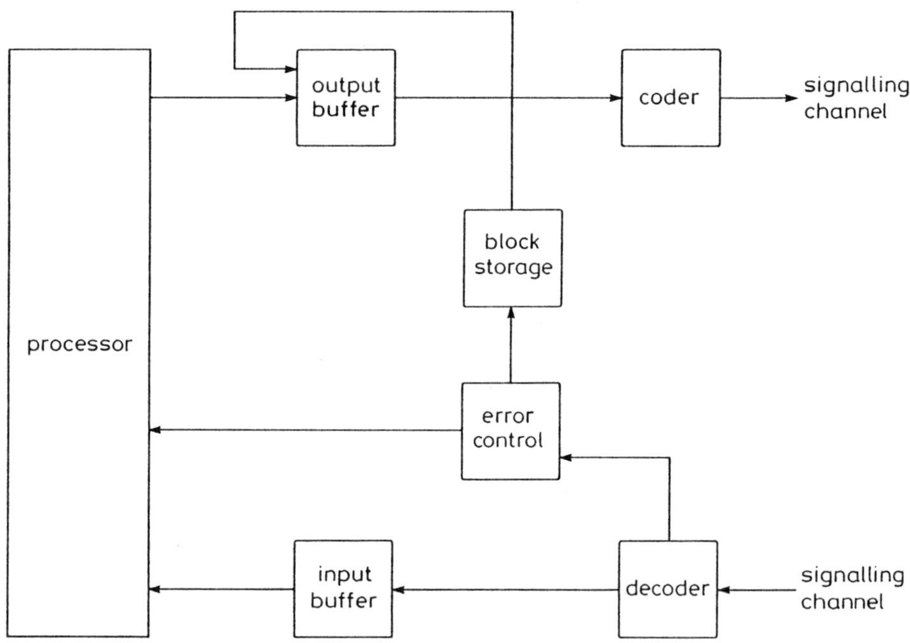

Fig. 3.1   Exchange functions for CCITT No. 6
Source: CCITT Recommendation Q.251

When receiving signal units, the decoder accepts the signal units from the demodulator (of an analogue transmission system) or from the digital transmission system. The decoder checks each signal unit for errors in conjunction with the 8-check bits. If errors are detected, the signal unit is discarded. If the signal unit is error-free, it is delivered to the input buffer, which stores the signal unit until the processor is available to analyse its contents.

Errors detected in signal units are overcome by requesting the retransmission of the corrupted signal unit. Thus, a copy of each transmitted signal unit is stored in 'block storage' until receipt of the signal unit by the other exchange is acknowledged. The acknowledgements are analysed in 'error control'. Successful acknowledgement allows deletion of the signal unit stored in block storage, whereas unsuccessful acknowledgement invokes the retransmission of the signal unit by the outgoing buffer.

CCITT No. 6 can be described in terms of:

(*a*)   The format of the signal units,
(*b*)   The procedures for sending and receiving signal units and
(*c*)   The error detection and correction techniques employed.

3.2 Formatting principles

3.2.1 General

Signalling information is transferred in one or more signal units, each having a fixed length of 28 bits. If the signalling information is contained within one signal unit, that signal unit is termed a 'lone-signal unit (LSU)'. LSUs are designed to carry very simple messages. If the signalling information is too large to fit within a LSU, then the information is carried in a series of signal units. The combination of signal units in this case is termed a 'multi-unit message (MUM)'. A MUM can consist of 2 to 6 signal units. The first signal unit in a MUM is termed the 'initial-signal unit'; the remaining signal units within a MUM are termed 'subsequent-signal units'. MUMs are designed to carry complex messages or messages containing a relatively-large amount of data (e.g. address information dialled by the calling customer).

3.2.2 Lone-signal unit

The basic format of the LSU is shown in Fig. 3.2. The format indicates how the bits within the signal unit are used to define the meaning of the message.

The heading code indicates the general class of signal being transmitted and the signal-information field distinguishes the particular signal within the class. Hence, the combination of the heading code and the signal-information field defines the specific signal being sent.

The label defines the speech circuit to which the message refers. The identification of the message by the number of the speech circuit to which the message refers indicates that CCITT No. 6 is circuit-related in nature. The 11 bits in the label allow up to 2048 speech circuits to be identified.

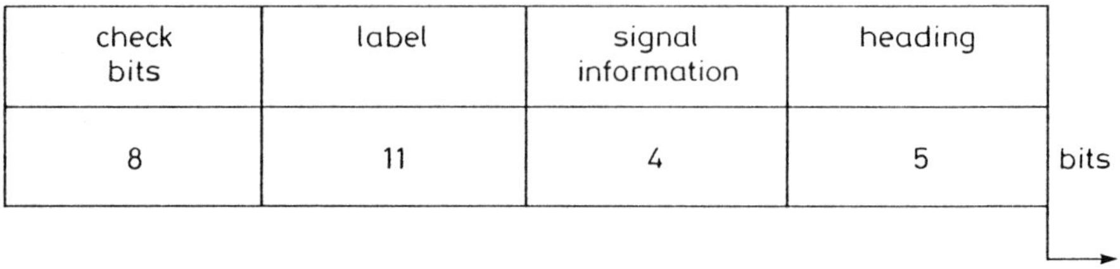

check bits	label	signal information	heading	
8	11	4	5	bits

Fig. 3.2   Format of the lone-signal unit

The check bits comprise the error-detection information generated by the coder to detect corruption of messages during transmission from one exchange to another.

Codings for some LSUs used during basic call control in telephony are shown in Table 3.1.

3.2.3 Multi-unit message

The basic format of an initial-signal unit within a MUM is the same as the LSU (shown in Fig. 3.2). In this case, the signal-information field identifies the signal unit as an initial-signal unit and indicates that subsequent-signal units will follow. Subsequent-signal units within a MUM have the general format shown in Fig. 3.3.

Table 3.1   Examples of codings for LSUs

Signal	Description	Heading Field	Signal-Information Field
Address Complete	Sent by the destination exchange to indicate that enough digits have been received to route a call to the called customer	11011	1010
Answer	Sent by the destination exchange to indicate that the called customer has answered the call	11000	0010
Clear Forward	Sent by the originating exchange to initiate the release of the call (eg. initiated by the calling customer ceasing the call)	11010	0010

check bits	signal information	length indicator	heading (00)	
8	16	2	2	bits

Fig. 3.3   Format of the subsequent-signal unit

The heading field is 2 bits long and is coded 00, thus easing the identification of the signal unit as a subsequent-signal unit. The length indicator is 2 bits long and it indicates the number of subsequent-signal units in the MUM, each subsequent-signal unit in a MUM carrying the same length indicator. A label is not required in a subsequent-signal unit, this information being contained in the LSU.

3.2.4 Initial-address message (IAM)

To illustrate the combination of an initial-signal unit with subsequent-signal units, forming a MUM, Fig. 3.4 shows the general format of an initial-address message (IAM). The IAM is the first message to be sent by an exchange to initiate the establishment of a call. The IAM is always a MUM due to the need to include routeing information and address digits. The information content is therefore too large to fit within a LSU. The IAM comprises a minimum of three signal units and a maximum of six signal units.

3.2.5 Signalling-system-control signals

In addition to signals required for telephony call control, CCITT No. 6 includes signalling-system-control signals to ensure that the signalling system itself functions correctly. Such signals are transferred as LSUs and include the 'acknowledgement' signal unit (ASU), the 'synchronisation' signal unit (SYU), the 'system-control' signal unit (SCU) and the 'multi-block-synchronisation' (MBS) signal unit.

check bits	circuit label	signal information	heading
x x x x x x x x	x x x x x x x x x x x	0 0 0 0	1 0 0 0 0

check bits	routeing information	length indicator	heading
x x x x x x x x	x x x x x x x x x x x x x x x x	01 *	0 0

check bits	address digits				length indicator	heading
	4th	3rd	2nd	1st		
x x x x x x x x	xxxx	xxxx	xxxx	xxxx	01 *	0 0

Fig. 3.4   Format of an initial-address message MUM

x indicates a bit of value 0 or 1

* code 01 indicates 2 subsequent-signal units in this example of the initial-address message

The ACU is used as part of the error-correction mechanism. The ACU indicates to an exchange having sent a block of signal units whether or not the block was corrupted when received by the other exchange. Each block contains 11 signal units plus an ACU-signal unit. The format of the ACU is illustrated in Fig. 3.5. The eleven acknowledgement indicators are used to define the status of the eleven signal units in the block that the ACU is acknowledging. A bit value of 0 indicates that the corresponding signal unit was received error-free, whilst a bit value of 1 indicates that an error was detected in the corresponding signal unit. The use of the ACU is described in Section 3.5.2. The block-sequence numbers are incremented cyclically with a modulo 8. The block-acknowledge-sequence number refers to the number of the block being acknowledged and the block-ACU-sequence number refers to the number of the block of which the ACU forms a part.

The SYU is used to maintain synchronisation of the signalling channel in the absence of signalling traffic. The format of the SYU is illustrated in Figure 3.6. The pattern was selected to help the synchronisation process. The block-SYU-sequence number refers to the position of the SYU within the block.

check bits	block–sequence numbers		acknowledgement indicators	heading
	ACU	acknowledge		
xxxxxxxx	x x x	x x x	x x x x x x x x x x x x	011

Fig. 3.5   Format of the acknowledgement-signal unit
x indicates a bit value of 0 or 1

check bits	block– SYU sequence number	synchronisation pattern
x x x x x x x x	x x x x	1110111 0111100011

Fig. 3.6   Format of the synchronisation-signal unit
x indicates a bit value of 0 or 1

The SCU is used to carry out control functions on the signalling link. For example, a 'changeover' signal indicates that a particular signalling link has failed and that is necessary to change to another signalling link to maintain service. The format of the SCU is illustrated in Fig. 3.7, the control-information field defining the function to be performed (e.g. changeover).

The original specification of CCITT No. 6 limited the number of blocks within the error-control loop to eight. This limit has subsequently been increased to 256, based on multi-blocks of 8 blocks. This gives rise to the need for a multi-block synchronisation (MBS) signal unit to enable the monitoring of multi-blocks. The format of the MBS is similar to the SCU.

check bits	control information	signal information	heading
x x x x x x x x	x x x x x x x x x x x x	1 1 0 0	11101

Fig. 3.7.   Format of the system-control signal unit
x indicates a bit value of 0 or 1

3.2.6 Management signals

Management signals relate to the management and maintenance of the signalling system and the speech network. Management signals are carried in LSUs or MUMs. The format of a management LSU is shown in Fig. 3.8. The initial-signal unit of a management MUM is similar to that shown in Fig. 3.8. The format of a subsequent-signal unit of a management MUM is shown in Fig. 3.9.

check bits	management information	signal information	heading
x x x x x x x x	x x x x x x x x x x x x	x x x x	11101

Fig. 3.8   Format of a management LSU
x indicates a bit value of 0 or 1

The specification of management signals [3] defines a range of codes. Examples are 'all circuits busy', which is used when there are no speech circuits available to the destination and 'switching-centre congestion', which is used to indicate that an exchange on the route to the destination cannot handle any more calls.

3.3 Procedures

Signalling system procedures define the sequences of message interchanges

check bits	management information	length indicator	heading
x x x x x x x x	x x x x x x x x x x x x x x x x	x x	0 0

Fig. 3.9   Format of a management SSU
x indicates a bit value of 0 or 1

to, for example, set-up and clear-down calls. The procedures for the set-up and release of a basic telephone call are shown in Fig. 3.10.

To initiate a call, the calling customer lifts the handset of the telephone and dials the address of the called customer. The originating exchange analyses the dialled digits received from the calling customer. When sufficient digits have been received to identify to which exchange the call needs to be routed, the originating exchange formulates an initial-address message and sends it to the next exchange (in this case the intermediate exchange). The initial-address message includes the digits dialled so far by the calling customer and also includes information on the type of speech path being provided e.g. whether or not a satellite link is being used as one of the transmission paths. The sending of an initial-address message automatically reserves an appropriate speech circuit for the call, the speech circuit being identified by the label. The initial-address message is analysed by the intermediate exchange to determine to which exchange the call should be routed. After analysis, an initial-address message is sent to the appropriate exchange (in this case the destination exchange). Additional address digits received by the originating exchange from the calling customer are passed to the destination exchange in 'subsequent-address' messages. When all address digits have been received at the destination exchange, the destination exchange returns an 'address-complete signal' and this is passed to the originating exchange. This message is sent to indicate to the originating and intermediate exchanges that enough address digits have been supplied to complete the call to the called customer. In some implementations, this allows certain call-control information stored in the originating and intermediate exchanges to be released from short-term memories, thus making the memories available for new calls. Ringing current is applied by the destination exchange to the called customer and ring tone is applied to the speech path by the destination exchange to inform the calling customer of the status of the call.

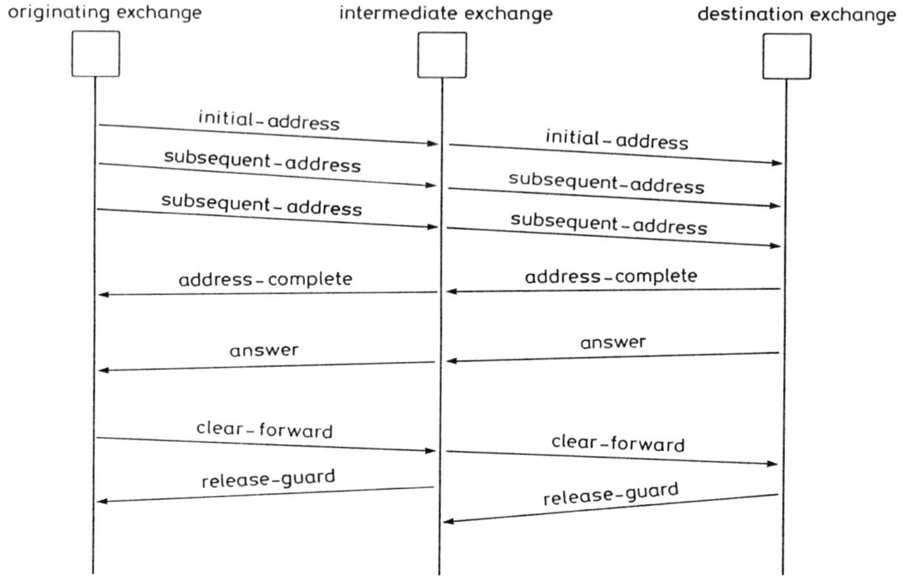

Fig 3.10   Procedures for a basic telephone call

When the called customer answers, an answer signal is sent from the destination exchange to the originating exchange via the intermediate exchange. In most networks, charging commences upon receipt of the answer signal at the exchange performing the charging function (typically the originating exchange). Conversation between the calling and called customers can now take place.

When the calling customer clears the call by replacing the handset, the on-hook condition is detected by the originating exchange, which sends a 'clear-forward' signal to the intermediate exchange. The intermediate exchange commences clear-down of the speech path and sends a clear-forward signal to the destination exchange. When clear-down of the speech path is completed at the intermediate exchange, the intermediate exchange returns a 'release-guard' signal to the originating exchange. Similarly, the destination exchange commences call clear-down upon receipt of the clear-forward signal from the intermediate exchange and returns a release-guard signal to the intermediate exchange when clear-down is complete. The originating and intermediate exchanges do not allow the reselection of the relevant speech path for another call until the release-guard signal has been received.

Thus, the release-guard sequence prevents a speech path being selected for a new call before that speech path has been fully cleared-down from a previous call.

3.4 Continuity check

In analogue CAS systems, the signalling path is coincident with the speech path being established for a call. In CCS systems, the physical tie between signalling and speech paths is removed. In early CCS systems, there was concern that the separation of signalling and speech paths could result in error conditions in which signalling messages were not associated with the correct speech path.

In response, CCITT No. 6 specifies the use of a 'continuity check', during call establishment, to ensure that there is a correct association of signalling and speech paths. This approach is not required in modern CCS systems due to the improved reliability of telecommunications equipment, but it can be provided as an option.

The continuity check involves sending a tone down the selected speech path and ensuring that the tone is received correctly. This process confirms that the speech path is coherent and that the correct signalling messages are associated with the speech path. The tone frequency specified by CCITT [4] is 2,000 Hz. If the continuity check for a call on one speech path fails, then an attempt is made automatically to establish the call on another speech path.

3.5 Error control

3.5.1 Error Detection

CCITT No. 6 signal units contain check bits for error-detection purposes. Referring to Fig. 3.1, the processor formulates signals that are 20 bits long. Each signal is passed to the output buffer to await its turn to be transmitted. When the turn of a particular signal arrives, the 20-bit signal is passed to the coder. The coder applies a specified polynomial [5] to the 20-bit signal, generates a corresponding 8-bit field and adds it to the signal. Thus, a signal unit transmitted from one exchange to another is 28 bits in length. Further details of the principles involved in applying a polynomial to generate check bits are given in Chapter 5. Upon receipt at the receiving exchange, the signal unit is passed to a decoder that removes the 8-bit field and applies the reverse process to the one adopted by the transmitting-exchange coder. If the application of the polynomial is successful, then it is assumed that the message has not been corrupted during transmission and it is passed to the input buffer. If the application is unsuccessful, then a transmission error is detected and the error-control function is informed. The approach adopted for CCITT No. 6 is optimised for detecting short noise bursts in the transmission system: longer noise bursts, or faults, are detected by circuit-failure mechanisms within the transmission system itself. Limiting the number of check bits to 8 reduces the overhead of the error-detection mechanism upon message-transmission capability, but the effectiveness of the error-detection capability is also restricted.

3.5.2 Error correction

Once an error is detected, it is necessary to correct the corrupted message. This achieved by retransmitting the message in question.

Consider two exchanges connected by CCITT No.6, as shown in Fig. 3.11. Assume that Exchange 1 has sent Block A to Exchange 2 and that a copy of Block A is kept in block storage of Exchange 1. Block A consists of 12 signal units, comprising 11 call-control signal units (Signal Units A.1 — A.11) and an ACU (Signal Unit A.12). After analysis at Exchange 2, it is recognised that all signal units except A.5 have been received without corruption. However, upon checking the 8-bit field in the decoder of Exchange 2, it is recognised that Signal Unit A.5 has been corrupted. In this case, the next available block of signal units to be transmitted from Exchange 2 to Exchange 1 is used to inform Exchange 1 of the corrupted Signal Unit A.5. In Figure 3.11, Block D is used for this purpose. Again, Block D consists of 12 signal units, with the 12th signal unit being an ACU. In the ACU, the acknowledgement indicators are set to indicate that Signal Units A.1 to A.4 and A.6 to A.11 were received successfully, whereas Signal Unit A.5 was corrupted. This information is analysed by the error-control function in Exchange 1, which informs block storage to eliminate the copy of the successfully received signal units and to re-transmit Signal Unit A.5. Thus, Signal Unit A.5 is passed to the output buffer to undergo retransmission to Exchange 2.

The error-control mechanism used in CCITT No. 6 is limited and can result in an incorrect sequence of messages, duplicated messages and unnecessary retransmissions. These abnormalities can particularly occur during fault conditions. For example, if a signal unit is received in error, a request to retransmit the signal unit is issued. However, it is possible that the next signal unit relating to the same call is received before the original signal unit is retransmitted. Thus, an exchange might receive signal units relating to a call in an incorrect order. To overcome some of these abnormalities, 'reasonableness' checks are incorporated in the signalling system. These require analysis within an exchange to determine whether or not certain message sequences are reasonable. This approach complicates the implementation of the signalling system and uses valuable processing power.

3.6 Comparison of CCITT Signalling Systems Nos. 6 and 7

It is useful to compare CCITT No. 6 with the more modern CCITT No. 7.

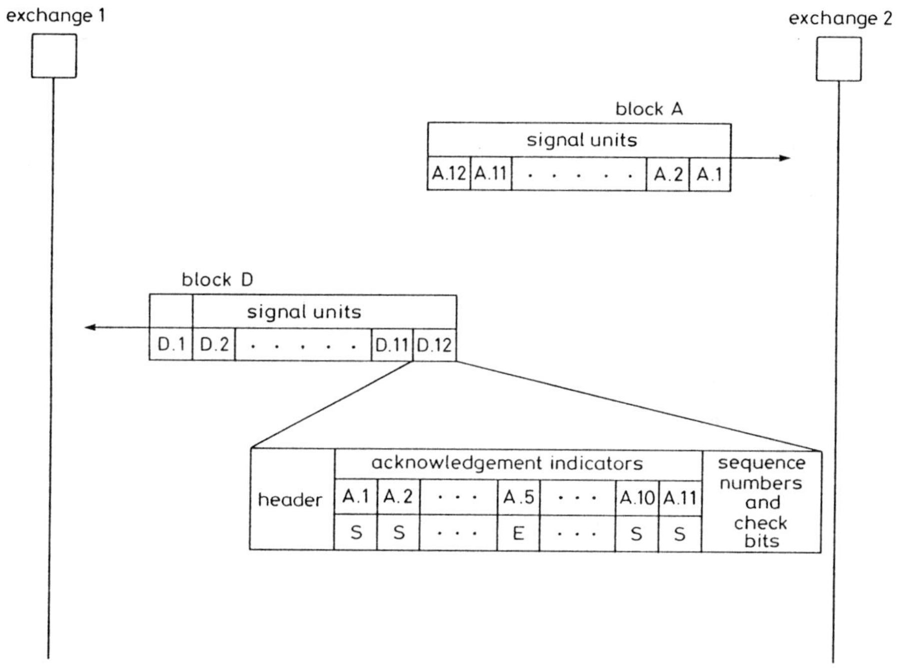

Fig. 3.11   Example of use of error-correction mechanism
S denotes successful transmission
E denotes error detected

Details of CCITT No. 7 are given in subsequent chapters.

CCITT No. 6 is a CCS system exhibiting many attributes that represent a quantum step above the capabilities of CAS systems. CCITT No. 6 handles the signalling information for numerous traffic circuits on a common path, allocating signalling capacity dynamically in response to demand. It is a message-based system, thus allowing a wide repertoire of signals and being compatible with modern-technology exchanges. CCITT No. 6 exhibits many network characteristics, allowing changeover from faulty signalling links, signal-transfer-point working and carrying network-management and maintenance information. It also provides a basic error-detection and error-correction mechanism to enhance reliability, albeit that the techniques adopted are not comprehensive.

However, despite all these attributes, CCITT No. 6 is being superseded by CCITT No. 7. The single most important reason for this is the limited degree of evolutionary potential incorporated within CCITT No. 6. Chapter 4 explains the structured approach being adopted for the specification of modern CCS systems (CCITT No. 7 and DSS1). This structured approach is the key to evolutionary potential, allowing parts of the signalling system to evolve without affecting other parts. CCITT No. 6 is not specified in this way, and the result is an automatic constraint on its ability to provide future flexibility. The formatting technique of CCITT No.6 is extremely limiting. Not only are signal units of 28 bits very small in the face of rapidly-increasing signalling-capacity requirements, but there is not an easy way to increase the size of the signal units. Modern signalling-system formats allow variable-length signal units. CCITT No. 6 is circuit-related in nature, whereas modern signalling systems need to provide both circuit-related and non-circuit-related capabilities. These non-circuit-related capabilities allow modern signalling systems to be able to provide signalling for all services, rather than being limited to, for example, telephony. Indeed, the combination of CCITT No. 7 and DSS1 provides a major step towards the objective of comprehensive and unimpeded signalling capability in telecommunications networks.

CCITT No. 6 was optimised for use in the analogue international telephone network. On the other hand, CCITT No. 7 is optimised for use in digital networks, both national and international, and it covers a range of services in a flexible and efficient manner. Hence, whilst CCITT No. 6 is used in the international network, and in a limited number of national networks, CCITT No. 7 is being introduced rapidly throughout the world in both national and international networks.

A summary of the key differences between CCITT Signalling Systems Nos. 6 and 7 is given in Table 3.2.

3.7 Chapter summary

CCITT No. 6 was the first CCS system to be implemented internationally. It exhibits many of the advantages of CCS systems, but it has limited evolutionary potential. This drawback is the main factor explaining why CCITT No. 6 is being superseded by CCITT No. 7.

Signalling information is included in signal units, which are 28 bits in length. Signal units are grouped into blocks of twelve. Message-signal units can either be lone-signal units (LSU) or multi-unit messages (MUM).

Lone signal units, and the first-signal unit of MUMs, consist of a heading code, signal information, a label and check bits. The combination of the heading code and signal information defines the meaning of the signal. The label defines the speech circuit to which the message refers. The check bits are used to check for corruption during transmission of the message.

Table 3.2 Comparison of CCITT Signalling Systems Nos. 6 and 7

Feature	CCITT No.6	CCITT No.7
Designed Environment	1960s	1970s, 1980s and 1990s
Environment	Optimised for analogue	Optimised for digital
Architecture	Unstructured, limiting evolutionary capability	Structured to allow evolution: aligns with structure adopted by international organisations
Application	Optimised for telephony, ie. set-up and release of speech paths	Covers telephony, circuit-switched data, ISDN, network features and general data-transfer mechanism
Signalling Transfer	Circuit-related	Both circuit-related and non-circuit-related
Coding Structure	Fixed-length signal units	Variable-length messages with evolutionary formatting
Error Control	Can lead to unsequenced and duplicated messages: needs reasonableness checks	Minimal lost or duplicated messages, even during fault conditions

Subsequent-signal units (SSUs) again contain a heading code and check bits. The rest of the SSU format consists of a length indicator and a signal-information field that is used to transfer, for example, address digits.

In addition to signal units for telephony, CCITT No.6 defines control signals and management signals that are used to operate and maintain the signalling system.

Calls are established by sending an initial-address message through the network. When sufficient address digits have been received by the destination exchange, an address-complete message is returned. This is followed by an answer message when the called customer answers the call. The call is cleared by using clear-forward and release-guard messages. Procedures are defined to conduct a continuity check of the speech path during the call-establishment procedure.

Errors are detected by monitoring the 8-check bits in each signal unit. Each signal unit is acknowledged as having been received successfully or unsuccessfully. If a message is detected as corrupted, it is retransmitted by the sending exchange.

Comparing CCITT No. 6 with CCITT No. 7 highlights the drawback of lack of evolutionary potential. The specification of CCITT No. 6 is not structured in a tiered approach, thus limiting the ability to modify the system. CCITT No. 6 is circuit-related only in nature and is optimised for telephony. The coding structure is fixed in length and the error-control mechanism is limited. These limitations of CCITT No. 6 are overcome in CCITT No. 7.

3.8 References

1 CCITT Recommendations Q.251 - Q.300: 'Specifications of Signalling System No. 6' (ITU, Geneva)
2 DAHLBOM, C A: 'Common channel signalling - a new flexible inter-office signalling technique', IEEE International Switching Symposium Record, Boston, 1972
3 CCITT Recommendation Q.260: 'Management signals' (ITU, Geneva)
4 CCITT Recommendation Q.271: 'Continuity check of the Speech path' (ITU, Geneva)
5 CCITT Recommendation Q.277: 'Error control' (ITU, Geneva)

Chapter 4

Architecture of modern CCS systems

4.1 Introduction

Modern signalling systems are very complex and they need to continue to evolve to match more demanding requirements of customers and networks. It is essential that modern signalling systems are flexible enough to handle new services and changes to existing services. To achieve this flexibility, modern CCS systems are specified in a structured way, thus allowing the specification to evolve in a controlled manner. A structured specification also allows a disciplined approach to design and development, thus easing the process of implementation. The term 'architecture' is used to describe the structured approach to the specification of CCS systems. The architecture of CCS systems is the key to their flexibility and evolutionary capability.

4.2 Requirements of architecture

A large number of factors influence the derivation of the architecture [1] of CCS systems. The important factors are given below.

(*a*)   The architecture of a CCS system must provide the potential for evolution. The signalling system must be able to handle future, as well as current, requirements. The implementation of new concepts should be achievable without major changes to the signalling system. It must be possible to continue the evolutionary process towards a general-purpose data-transfer network, allowing unimpeded communication between customers, between customers and the network and between nodes within the network.

(*b*)   A modern CCS system in a network comprising software-controlled nodes is very complex. It must be possible to optimise one aspect of the signalling system without affecting other parts. If this is not achieved, the flexibility of the signalling system is seriously curtailed: flexibility is a key requirement.

(*c*)   The architecture should allow the signalling system to be used for a range of applications. In the past, internationally-specified signalling systems have only catered for telephony. Modern signalling systems need to cater for a variety of services including telephony, data, ISDN and mobile services. Network features must also be covered, including network management and operations and maintenance aspects.

(*d*)   As signalling networks become progressively more important, it is essential that reliability of communication is adequately covered. Loss of a signalling link can have a major impact upon a large number of customers. The architecture of the signalling system must allow an effective way of ensuring high reliability.

(*e*)   It must be possible to make changes to existing services, e.g. in response to customers' reactions to service provision. It must also be possible to add new services and features. CCS systems must be able to match the changing requirements placed upon the network. This continual up-dating process results in new versions of CCS systems being imple-mented. These new versions must be able to interwork with previous versions that have already been implemented. The architecture should provide the ability to ease this process.

(*f*)   In the past, internationally-specified signalling systems were designed for use in the international network, with only a cursory consideration of national-network requirements. This approach resulted in expen-sive signalling interworking requirements at the boundary of the international network with each national network. One aim of a structured approach to specification is to develop signalling systems for use in both national and international networks, thus reducing the interworking costs. In addition, national-signalling systems developed to international specifications reduce the overall implementation cost due to economies of scale: both CCITT No. 7 and DSS1 benefit from worldwide application.

(*g*)   The specification of the signalling system needs to be sufficiently comprehensive to allow different exchange manufacturers to develop implementations independently. However, the specification must achieve satisfactory interworking between different exchange systems. If the signalling system is defined in too much detail, the innovation of the exchange manufacturer is constrained. If not enough detail is given, different implementations will not interwork. The correct balance is difficult to achieve, but it is a key to cost-effective implemen-tation. One reason for this difficulty is the relationship between exchange call-control and CCS procedures. There is a strong link between these two issues : changing one can have a great impact upon the other. A means of specification is required to achieve the correct balance.

(*h*)   Similarly, the interworking of networks must be considered. When it is required for one network operator to interconnect with another, the signalling system must be defined sufficiently to allow interworking to take place. However, it must also allow each network to operate and evolve independently.

(*i*)   The specification of a modern signalling system must be conducted in a disciplined and logical manner. Not only must each feature be specified comprehensively, but procedures to handle any failure conditions must be rigorously defined. Each element of the specification must be defined as clearly as possible, reducing ambiguity to a minimum. The architecture of the signalling system must help the specification process to achieve a clear and unambiguous definition.

(*j*)   The signalling system must be able to interwork with equipment already existing in the network. National networks in the past tended to develop independently in the absence of standards agreed between countries. Hence, numerous national-signalling systems exist and these must be able to interwork satisfactorily with evolving CCS systems.

(*k*)   In the past, standards for signalling systems have taken a great deal of time to develop. This has resulted in network operators developing equipment to interim standards to meet customer needs. This approach negates the advantage of economies of scale that applies to a system designed to an international standard. It also results in reluctance to derive new standards because there is a wide variety of conditions from which networks operators need to evolve, rather than a common platform based on one standard. The architecture must therefore allow the speedy specification of evolving standards and new standards. A structured architecture, allowing one part of the system to be modified without affecting other parts, is essential to facilitate speedy specification.

4.3 Architecture development

The requirements listed above necessitate a structured definition of CCS systems. Modern CCS systems are defined in a series of 'tiers'. Consider two exchanges needing to transfer information, as shown in Fig. 4.1. The signalling functions at each exchange can be categorised into Tiers X, Y and Z. The concept is that each tier contains a prescribed set of functions. The interface between tiers is closely defined and information can only pass between tiers by the use of units of information specified for that purpose, termed 'primitives'. Primitives are defined in terms of information content, the electrical and physical characteristics of the interface being implementation-dependent. Each tier is self-contained and can be optimised and/or modified without affecting other tiers, provided that the interface between tiers remains constant (i.e. the primitives are not changed). Each tier, in conjunction with the lower tiers, provides a 'service' to the tier above. For example, the functions in Tier X of Exchange A determine the physical and electrical conditions that apply to the transmission link. Tier Y of Exchange A can assume that the details of the transmission link are handled by Tier X: Tier Y can thus concentrate on providing other functions for Tier Z. The functions within a tier at Exchange A can communicate with the functions of the same tier at Exchange B by making use of the lower tiers. The specification of CCS systems is achieved by defining:

(*a*) The primitives used between adjacent tiers within a node,

(*b*) How the information passed from a tier in one node to a corresponding tier in another node is coded (known as the 'format') and

(*c*) The logical sequence of information transfer between a tier in one node and a corresponding tier in another node (known as 'procedures').

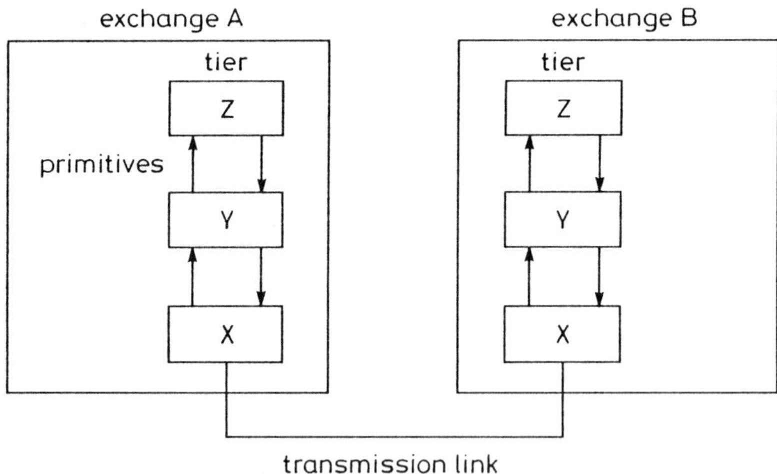

Fig. 4.1   Illustration of tier structure

The combination of items (*a*) to (*c*) is known as a 'protocol'. To illustrate the use of the tier structure, assume that Exchange A in Figure 4.1 needs to release a call that has been established between Exchanges A and B. Tier Z in Exchange A is responsible for recognising the need to release the call. Thus, Tier Z formulates appropriate information and passes it, using a primitive, to Tier Y. Tier Y is responsible for routeing a CCS message from Exchange

A to Exchange B. Thus Tier Y amalgamates the primitive from Tier Z with its own routeing information and passes the result to Tier X, again using a primitive. Tier X is responsible for selecting an appropriate transmission link and transferring the information from Tiers Z and Y to Exchange B in a CCS message. On receipt of the message, the tiers in Exchange B perform corresponding functions to those at Exchange A, resulting in Tier Y passing a primitive to Tier Z requesting the release of the call.

The tiered approach to specifying CCS systems allows each of the criteria in Section 4.2 to be met. By making each tier a self-contained group of functions, with defined responsibilities, one tier of the signalling system can be modified without affecting other tiers.

In practice, there are two types of tier structure used for specifying CCS systems. The original tier structure was adopted for CCITT No. 7 when the main applications for CCS systems were circuit-related. This structure [2] consists of four 'levels' and it is described in Section 4.4.

Two factors have arisen since the 4-level structure of CCITT No. 7 was derived. The first factor is that the International Standards Organisation (ISO) has developed a 7-layer open-systems-interconnection (OSI) protocol reference model [3, 4, 5]. Consisting of 7 'layers', rather than the original 4 levels for CCITT No. 7, the OSI model represents a more comprehensive and disciplined approach to architecture. The OSI model is likely to be used widely in future networks to rationalise approaches to protocol structure. The relationship between the OSI 7-layer model and the CCITT 4-level structure is explained in Sections 4.6 and 4.7. However, it is essential that the terminology is understood at this stage. The term 'level' refers to the original (circuit-related) structure for CCITT No. 7 and the term 'layer' refers to the OSI 7-layer model.

The second factor influencing the architecture is the need for CCS systems to be capable of non-circuit-related operation. An example of this type of communication is when there is a need to gain access to remote network databases storing specialised information. In this example, signalling is required in the absence of speech paths. The original 4-level structure is not as useful for non-circuit-related applications and the OSI 7-layer model is used in these cases.

Both types of tier structure apply to CCITT No. 7, with parts being defined in terms of the 4-level structure and other parts being defined in terms of the 7-layer model. DSS1 is defined only in the context of the 7-layer model. The 7-layer model is explained in Section 4.5 and its application to CCITT No. 7 is explained in Sections 4.6 and 4.7. The application of the 7-layer model to DSS1 is described in Section 4.8.

4.4 Level structure of CCITT No. 7

The requirements listed in Section 4.2 led to CCITT No. 7 being developed as a CCS system for use in both national and international networks. It is optimised for use in a digital environment, but it can be used on any transmission medium. CCITT No. 7 is highly flexible, facilitates the evolutionary process and supports a variety of services and network features. These attributes result from the early decision to specify CCITT No. 7 in a 4-level structure [6], as illustrated in Figure 4.2.

A prime objective when formulating the design of CCITT No. 7 was to ensure that the signalling system flexibly handled the requirements for circuit-related applications. These applications include telephony and circuit-switched data (i.e. data using circuits within transmission links in a similar way to telephone calls). The functions performed by the 4-level model are described in Sections 4.4.1 to 4.4.5, and Section 4.4.6 illustrates their application.

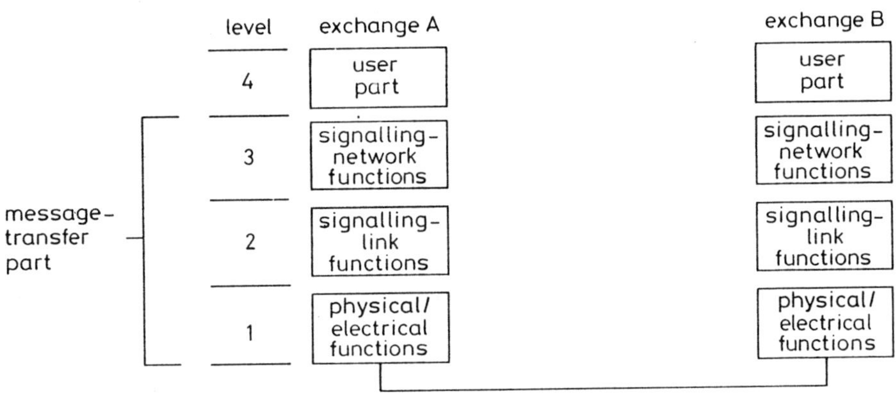

Fig. 4.2   Level structure of CCITT No.7

4.4.1 Level 1

Any node with the capability of handling CCS is termed a 'signalling point'. The direct interconnection of two signalling points with CCITT No. 7 uses one or more 'signalling link(s)'. Level 1 of the 4-level structure (shown in Fig. 4.2) defines the physical, electrical and functional characteristics of the signalling link. Defining such characteristics within Level 1 means that the rest of the signalling system (Levels 2 to 4) can be independent of the transmission medium adopted. By keeping the interface between Levels 1 and 2 constant, any changes within Level 1 do not affect the higher levels. In a digital environment, the usual physical link is a 64 kbit/s channel. This is typically within a digital transmission system using pulse-code modulation

(PCM). However, other types of link (including analogue) can be used without affecting Levels 2 to 4.

4.4.2 Level 2

Level 2 defines the functions that are relevant to an individual signalling link, including error control and link monitoring. Thus, Level 2 is responsible for the reliable transfer of signalling information between two directly-connected signalling points. If errors occur during transmission of the signalling information, it is the responsibility of Level 2 to invoke procedures to correct the errors. Such characteristics can be optimised without affecting the rest of the signalling system, provided that the interfaces to Levels 1 and 3 remain constant.

4.4.3 Level 3

The functions that are common to more than one signalling link, i.e. signalling-network functions, are defined in Level 3: these include 'message-handling' functions and 'signalling-network- management' functions. When a message is transferred between two exchanges, there are usually several routes that the message can take, including via a signal-transfer point. The message-handling functions are responsible for the routeing of the mes-sages through the signalling network to the correct exchange. Signalling-network-management functions control the configuration of the signalling network. These functions include network reconfigurations in response to status changes in the network. For example, if an exchange within the signalling network fails, the Level 3 of CCITT No.7 can re-route messages and avoid the exchange that has failed.

4.4.4 Message-transfer part

Levels 1 to 3 constitute a transfer mechanism that is responsible for transferring information in messages from one signalling point to another. The combination of Levels 1 to 3 is known as the message-transfer part (MTP). The MTP does not understand the meaning of the messages being transferred, but it controls a number of signalling-message, link and network-management functions to ensure correct delivery of messages. This means that the messages are delivered to the appropriate exchange in an uncorrupted form and in the sequence that they were sent, even under failure conditions in the network. Details of the MTP are given in Chapter 5.

4.4.5 Level 4

Level 4 comprises the 'user parts'. These define the meaning of the messages transferred by the MTP and the sequences of actions for a particular application (e.g. telephony). A key feature is that many different user parts may use the standardised MTP, as shown in Fig. 4.3. Hence, if new requirements arise, that had not been foreseen previously, the relevant user part can be enhanced (or a new user part derived) without modifying the transfer mechanism or affecting other user parts. Three user parts have been defined: the telephone user part (TUP), the ISDN user part (ISUP) and the data user part (DUP). The user parts are specified in terms of message formats and procedures. The message formats define the meaning of a particular message and specify the codings to be used. The procedures define the sequence of messages to be followed for the particular application (e.g. telephony). Hence, the messages to be exchanged and the sequences to be followed are defined for each application.

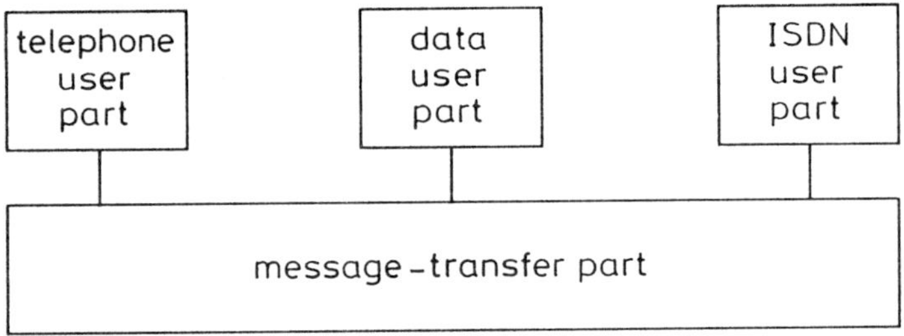

Fig. 4.3   Examples of multiple user parts

Whilst CCITT No. 7 is specified as a signalling system, Level 4 specifies a number of call-control functions. Indeed, the circuit-related mode of CCITT No. 7 is so closely associated with controlling the set-up and release of physical circuits that it is essential that some aspects of call-control are defined within the user part specification in order to optimise the procedures that are adopted. Details of user parts are given in Chapter 6.

4.4.6 Application of the level structure

The application of the level structure is illustrated in Fig. 4.4. Exchanges A and B are directly connected by speech circuits (denoted by the solid lines connecting the respective switch blocks). A signalling link is also available between Exchanges A and B (denoted by the dotted line). It is shown that Level 4 (the user part) is closely associated with the control function of the exchange.

If the control function of Exchange A needs to communicate with the control function of Exchange B (e.g. to initiate the set-up of a speech circuit between the exchanges), the control function in Exchange A requests the Level 4 functions to formulate an appropriate message. Level 4 then requests the message-transfer part (Levels 1 to 3) to transport the message to

Exchange B. Level 3 analyses the request and determines the means of routeing the message to Exchange B. The message is then transported via Levels 1 and 2.

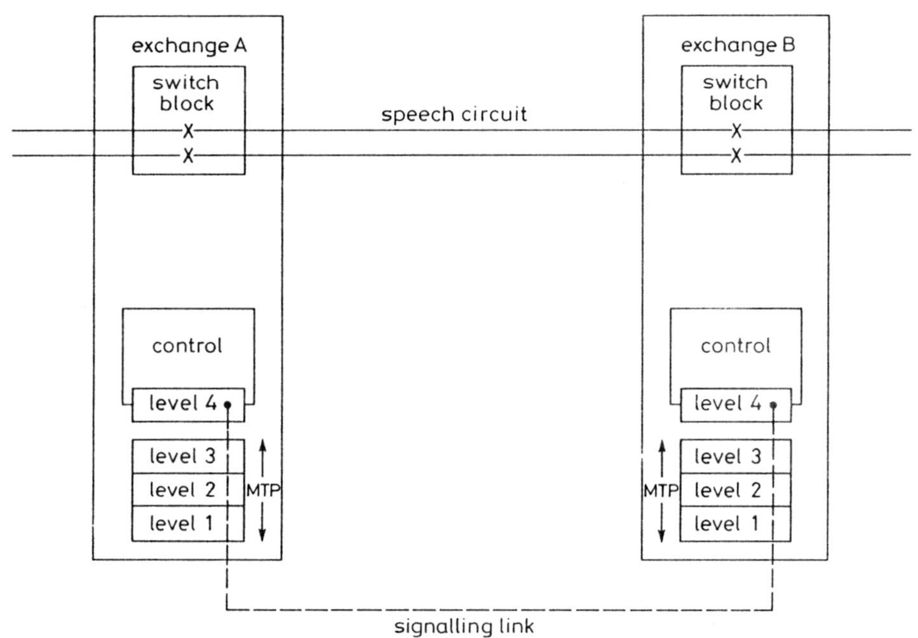

Fig. 4.4   Application of the level structure

Upon receipt of the message by the MTP of Exchange B, Levels 1 and 2 deliver the message to Level 3. Level 3 at Exchange B recognises that the message has arrived at the correct exchange and passes the message to Level 4. Level 4 in Exchange B then interacts with the control function to determine the appropriate action and response. If problems arise in the transmission process between Exchanges A and B, causing message corruption, the Level 2 functions are responsible for detecting the corruption and retransmitting the information. If the signalling link between Exchanges A and B is not available (e.g. failure of the link), the Level 3 functions are responsible for re-routeing the information through the signalling network to Exchange B.

Using these techniques, Exchanges A and B can send each other appropriate messages until the need to communicate on a particular transaction ceases (e.g. a speech circuit between exchanges is released).

Fig. 4.5 shows the more general case of Exchanges A and B communicating using quasi-associated signalling. In this case, Exchanges A and B are directly connected by speech circuits but the signalling links are available between Exchanges A and C and Exchanges C and B. In this case, Exchange C acts as a signal-transfer point (STP) in order to route signalling messages between Exchanges A and B. Hence, the MTP in Exchange A routes the message to Exchange C. As part of the signal-transfer point function, the Level 3 in Exchange C recognises that the message needs to be onward routed to Exchange B. Hence, Level 4 in Exchange C is not involved in the message transfer, thus avoiding Level-4 processing.

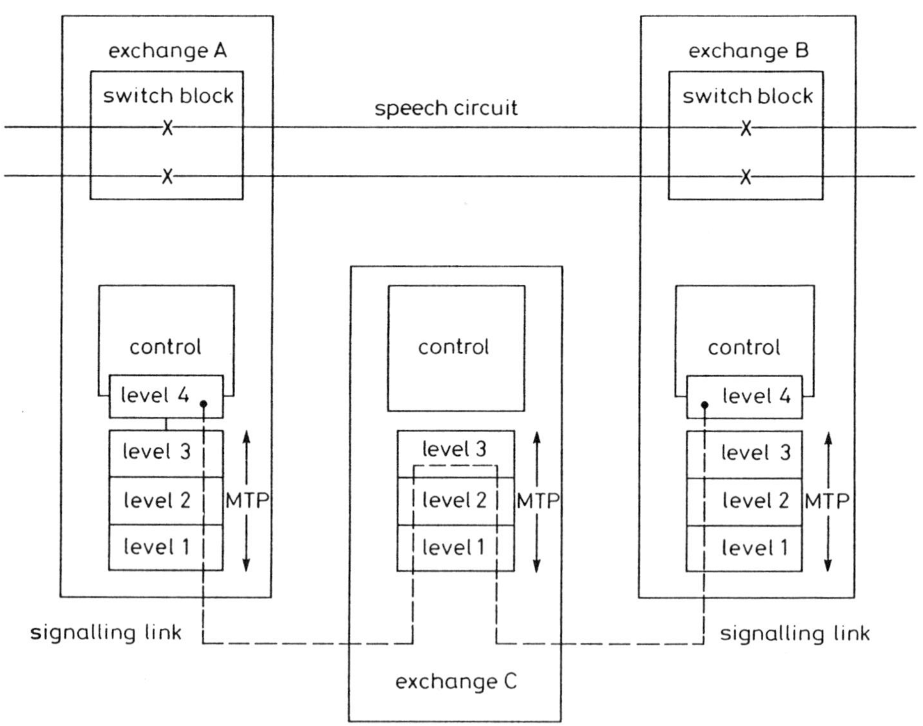

Fig. 4.5   Application of the level structure: signal-transfer point

4.5 The OSI 7-layer model

4.5.1 General

The OSI 7-layer model was derived with inter-processor data communication as the main field of application. The entities needing to communicate, termed 'users', were computers. The aim was to define the functions that need to be performed to allow communication between two computers in terms of 7 layers. However, the 7-layer model can also be applied in principle to other types of user, e.g. customers making telephone calls.

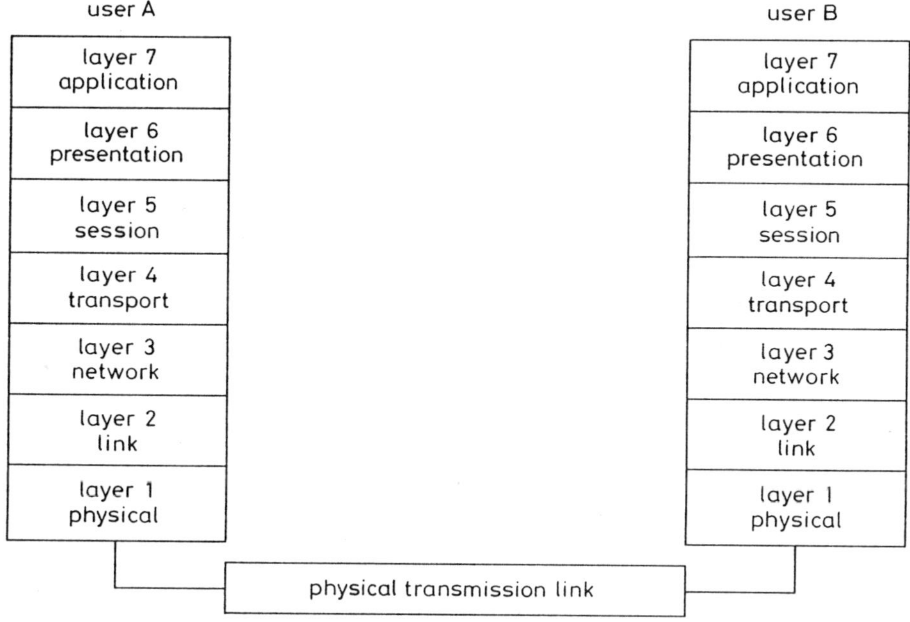

Fig. 4.6   OSI 7-layer model

The OSI model is shown in Fig. 4.6. In this context, the term user applies to customers' premises as well as network nodes (e.g. exchanges). The concept of the 7-layer model is similar to that adopted in the 4-level structure of CCITT No. 7. However, in this case, functions are categorised into 7 layers, each layer in User A being able to communicate with the corresponding layer in User B via the lower layers. Formats and procedures are defined to allow corresponding layers to communicate. Each of the 7 layers performs a defined set of functions. The functions in a particular layer, together with the functions provided by lower layers, are deemed to provide a 'service' to the layer above. For example, the network layer (Layer 3) is deemed to provide a network service to the higher layers.

Within a user, each layer can communicate with adjacent layers using primitives. Four types of primitive are used in the OSI 7-layer model. 'Request' and 'Indication' are primitives from one layer that request another layer to perform a function. 'Response' and 'Confirmation' are used to indicate that the function has been performed. An illustration of the use of the primitives is given in Section 4.5.9. Each of the primitives contains information to qualify the type, e.g. an 'establish- request' primitive is used to indicate that a signalling connection is required.

The 7 layers are supported by a physical transmission link that forms part of the telecommunications network, as shown in Fig. 4.6. The physical link can be any form of transmission system. Variations in transmission characteristics are handled by the Layer 1 functions.

The 7 layers in the model are described in Sections 4.5.2 to 4.5.8 below. Layers 1 to 4 deal with the establishment of a communication link between users whereas Layers 5 to 7 deal with communication over the link. In an OSI environment, a user is a person or computer needing to communicate with another person or computer. For telephony, a user can be a customer wishing to talk to another customer or it can be an exchange needing to communicate with another exchange (see Section 4.6).

4.5.2 Layer 1

Layer 1 (physical layer) relates to the physical transfer of a bit stream over a transmission medium. It provides the interface to the transmission medium. A 64 kbit/s link is the normal Layer 1 for CCS systems, but other transmission media are also possible.

4.5.3 Layer 2

Layer 2 (data-link layer) is responsible for overcoming deficiencies in the transmission medium by, for example, the provision of error detection and correction techniques. Hence, errors in transmission (e.g. from bursts of noise) causing corruption of messages are overcome. Layer 2 applies between two directly- connected nodes.

4.5.4 Layer 3

Layer 3 (network layer) transfers data through a network from one user to another user and is responsible for analysing address digits and routeing accordingly.

Layer 3 communication can either be 'connection-oriented' or 'connectionless'. In the connection-oriented case, a relationship needs to be established between two users to ensure co-ordination of the data that is exchanged. For example, consider the case that the users request the network to guarantee that messages sent by one user are delivered in the same order to the other user. In this case, a relationship between the two users needs to be established within the network to ensure that messages are not delivered out-of-sequence. Hence, a connection-oriented service is required.

The term connectionless is used when no guarantee is given that messages will have a particular relationship (e.g. delivery of messages in the order of sending them is not guaranteed by the network layer).

4.5.5 Layer 4

Layer 4 (transport) provides the ability to establish a transparent transfer of information from one user to another, relieving the users of involvement in the means of achieving the data transfer. Hence, Layer 4 allows the establishment of direct communication between two users. Layer 4 can also be used to enhance some of the functions of Layer 3, e.g. to provide a higher quality of service than normally provided by Layer 3.

4.5.6 Layer 5

The session layer is the lowest layer that deals with direct communication between the two users, as opposed to the establishment of the communication link. The session layer defines the type of interaction to be used between the two users, including the nature and timing of interactions. For example, the communication could be two-way simultaneous (i.e. both users able to communicate simultaneously), two-way alternate (i.e. both users able to communicate but one at a time) or one-way (only one user able to send information).

4.5.7 Layer 6

One of the aims of the OSI model is to allow users to adopt different data syntaxes and still be able to communicate with each other. Layer 6 (presentation) translates the syntax of the data being transferred between that used by one user and that used by the other user. If both users adopt the same syntax, this function is not required.

4.5.8 Layer 7

The application layer provides the interface between the communications environment and the user. Layer 7 lays down the type of communication required to satisfy the needs of the user.

4.5.9 Application of the layer structure

The general application of the 7-layer model is illustrated in Fig. 4.7. If Layer 7 in User A needs the corresponding layer in User B to perform an action, the request to perform the action is passed to Layer 6 in User A as a 'request' primitive. An example of such a primitive would be a request to perform a translation on a telephone number to allow a specialised routeing to occur. The primitive can include parameters to give additional information (e.g. the telephone number for which a translation is required). This information is passed through the lower layers at User A (Layers 1–5 are amalgamated in the diagram for clarity), across the transmission medium and through the lower layers at User B. Layer 6 at User B presents the request to perform the action as an 'indication' primitive to Layer 7. The answer to the request (e.g. stating that the translation has been performed) is passed to Layer 6 at User B as a 'response' primitive. This information is transferred to Layer 6 at User A using the lower layers and the transmission medium. Layer 6 at User A then presents the answer to Layer 7 as a 'confirmation' primitive. Parameters are associated with each primitive in the example and the confirmation primitive includes the translated number as a parameter.

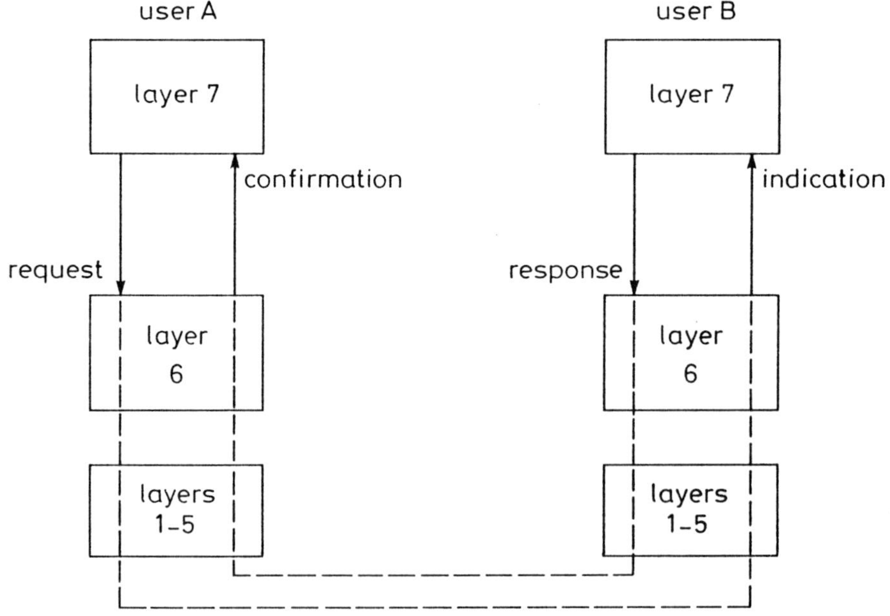

Fig. 4.7   Application of the 7-layer model

4.6 Application of the OSI model to CCITT No. 7

The application of the OSI model to CCITT No. 7 needs to take account of the context in which the model was developed. In the original OSI model, the users were computers, the telecommunications network providing the functions of Layers 1 to 3. The functions of Layers 4 to 7, in this case, are provided by the user. When the model is applied to a telephone network on a customer-to-customer basis, the same principles apply, with the telephone network providing the functions of Layers 1 to 3 and the customers (together with their telephones) providing the functions of Layers 4 to 7 . This is illustrated in Fig. 4.8, in which Exchanges A and B provide the functions of Layers 1 to 3 from a customer perspective. However, the categorisation of network functions into Layers 1 to 3 is not very helpful from a CCITT No.7 viewpoint. The use of only 3 layers does not provide a comprehensive enough structure on which to define CCITT No.7 non-circuit-related functions. Hence, for CCITT No.7 specification purposes, the OSI model is applied between Exchanges A and B, thus allowing the full 7 layers to be used. In this case, the users can be considered to be the two communicating processors in Exchanges A and B. This is illustrated in Fig.

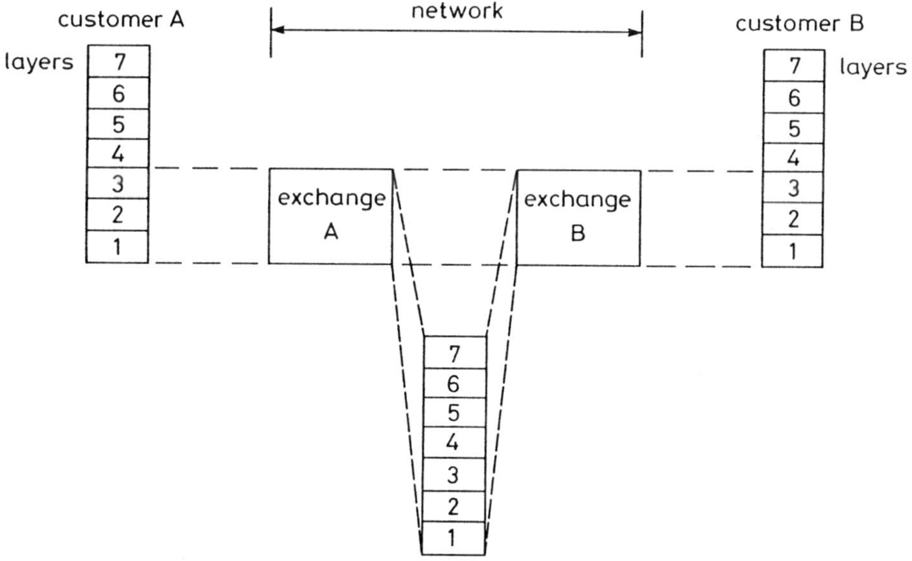

Fig. 4.8   Context of the OSI 7-layer model

4.8 by showing that the full 7 layers exist between Exchanges A and B when considered from a network viewpoint.

The result is that customers see the network as providing the functions of Layers 1–3. However, between the two exchanges (A and B), the full 7 layers of the OSI model can be applied.

The application of the 7-layer model to CCITT No.7 identifies two points. The first point is that, whilst the MTP (Levels 1 to 3 of the 4-level structure) provides a comprehensive transfer technique, it does not provide all the functions necessary to fulfil an OSI network service (i.e. appropriate functions of Layers 1 to 3 of the 7-layer model). Hence, a functional element, in addition to the MTP, is required to provide a network service. This element is called the 'signalling-connection-control part' (SCCP), and it is shown in Fig. 4.9. The SCCP is designed to enhance the MTP such that the transfer mechanism for CCITT No. 7 can meet the OSI Layer 3/4 boundary

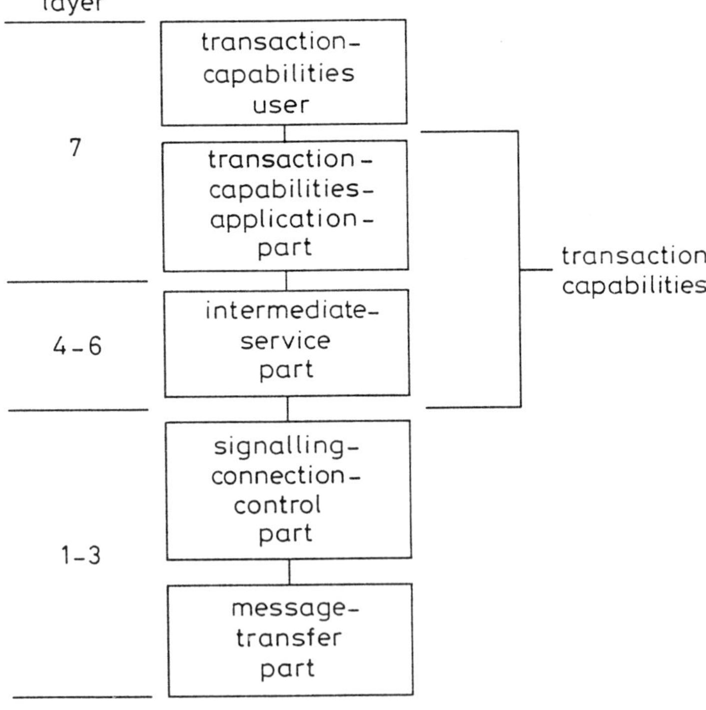

Fig. 4.9 Layer structure of CCITT No. 7
Source: CCITT Recommendation Q.771

requirements. Thus, the SCCP allows CCITT No. 7 to offer an OSI network service. Details of the SCCP are given in Chapter 5.

The second point that arises when applying the 7-layer model to CCITT No. 7 is that the user parts are large and complex functional elements, thus restricting the full benefits of a structured architecture. The telephone user part (TUP) , ISDN user part (ISUP) and data user part (DUP) were specified before the widespread adoption of the OSI 7-layer model and there is little benefit in re-specifying them to align with the model. However, non-circuit-related applications can be specified to align with the 7-layer model and this approach has led to the specification of a functional element termed 'Transaction Capabilities' (TC). This is shown in Fig. 4.9. By applying the OSI 7-layer model to TC, maximum advantage of a structured architecture is achieved. TC is a general protocol encompassing Layers 4 to 7 of the OSI model. TC is divided into the 'transaction-capability-application part' (TCAP) and the 'intermediate-service part' (ISP). TCAP, together with some functions provided by the user of TC, provides the functions of Layer 7. An example of a user of TC is the operations, maintenance and administration part (OMAP) of CCITT No. 7, further details of which are given in Chapter 7. The ISP provides the functions of Layers 4 to 6.

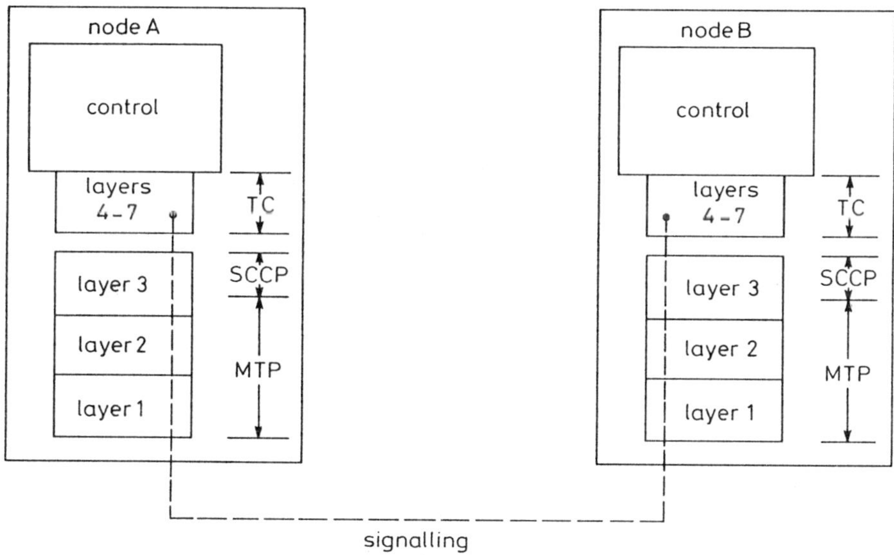

Fig. 4.10   Application of layered structure to CCITT No.7

The TC protocol is based on data-transfer protocols specified by CCITT[7], modified as necessary to meet the demands placed on CCITT No. 7. One concept behind TC is that it should be 'portable', i.e. TC should not only be able to use the functions provided by the MTP and SCCP, but also any other transport mechanism offering a network service. In a similar way, many different types of user should be able to use TC to establish communication with other users.

The application of the layered structure, in the context of exchange-to-exchange communication, is shown in Fig. 4.10. In contrast to Fig. 4.4, which illustrates the application of the level structure, Fig. 4.10 shows no speech circuits: the SCCP and TC operate in isolation from, and without the need for, speech circuits. The dotted line indicates a signalling connection. In this case, the control function is the user of TC. The SCCP forms part of the transfer mechanism. TC uses the SCCP and MTP to transfer information between Nodes A and B.

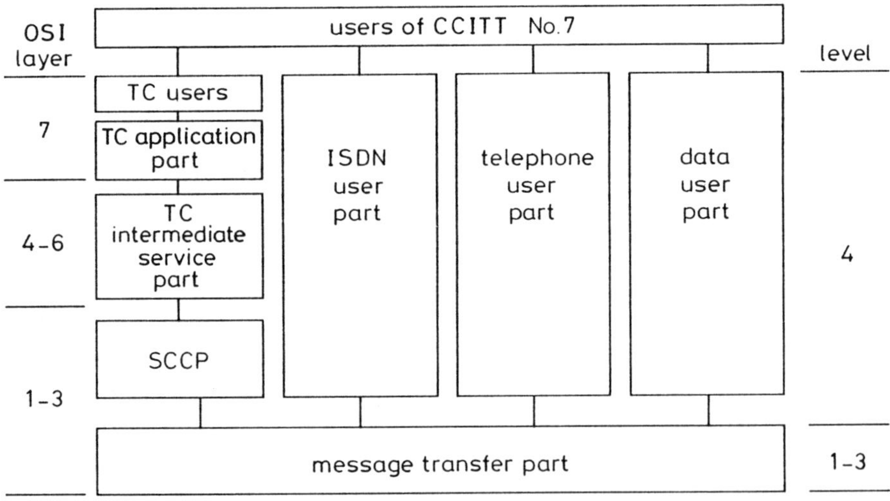

Fig. 4.11   Overall architecture of CCITT No.7

4.7 Overall architecture of CCITT No. 7

The overall architecture of CCITT No. 7 is shown in Fig. 4.11. It is built up from the original 4-level structure and the OSI 7-layer structure.

The MTP and SCCP are equivalent to Layers 1 to 3 of the 7-layer model and TC forms Layers 4 to 7. The user parts are equivalent to Layers 4 to 7 and they also contain some functions of Layer 3. The MTP, TUP and DUP will not be changed to meet the later OSI hierarchy. Future evolution of these elements is likely to be very limited. Hence, the advantages of a more-structured specification would not outweigh the difficulties of changing the specification after implementation. The ISDN user part (ISUP) needs to handle future services, e.g. controlling multiples of 64 kbit/s speech connections. Hence, further evolution of the ISDN user part is taking place. One concept for the ISDN user part is to split the user-part functions into a 'call-control part' and a 'bearer-connection-control part'. This approach breaks away from the concept that ISDN calls are necessarily circuit related. The call-control part handles those aspects of the user part that relate purely to the call, thus being independent of the traffic circuit being used. The bearer-connection-control part handles the functions for controlling corres-ponding traffic circuits (i.e. the circuit-related functions). Whilst not amounting to alignment with the 7-layer model, this approach reaps the benefits of a more-rigorous structured specification approach. The resulting user part is termed the 'ISDN Signalling-Control Part (ISCP)'.

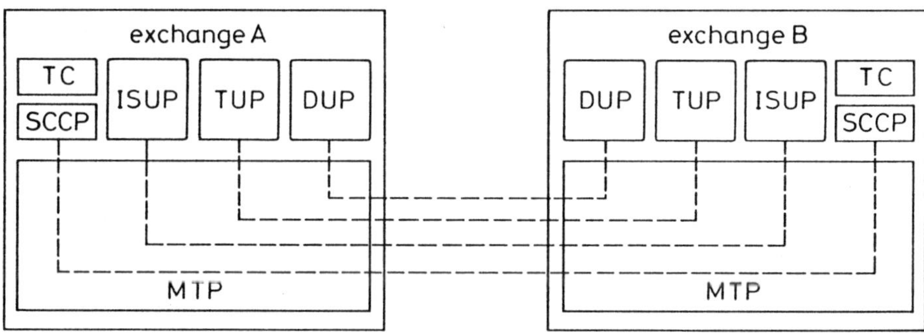

Fig. 4.12   Application of the overall architecture (inter-exchange)

The application of the overall architecture in an inter-exchange context is illustrated in Fig. 4.12. An exchange, in practice, does not provide the full range of user parts. A typical exchange has one user part (e.g. ISUP) and a non-circuit-related capability (TC and SCCP). The ISUP allows the exchange to set up and release traffic circuits for speech and circuit-switched data. TC and SCCP allow the exchange to invoke non-circuit-related functions when required. These features would, for example, allow the exchange to gain access to a remote network database for specialised routeing/translation information. Nodes like databases are provided with TC and SCCP for non-circuit-related transactions and a user part is not required, as illustrated in Fig. 4.13. In this Figure, two exchanges are interconnected by traffic circuits (denoted by solid lines). Consider an example of the ISUP of Exchange A receiving a request to set up a speech circuit. Exchange A recognises that the call requires specialised routeing information. Hence, the control function of Exchange A requests TC to determine the required routeing translation from the remote network database. This is performed using the MTP and SCCP and without setting up a speech circuit to the database. The translator in the database provides

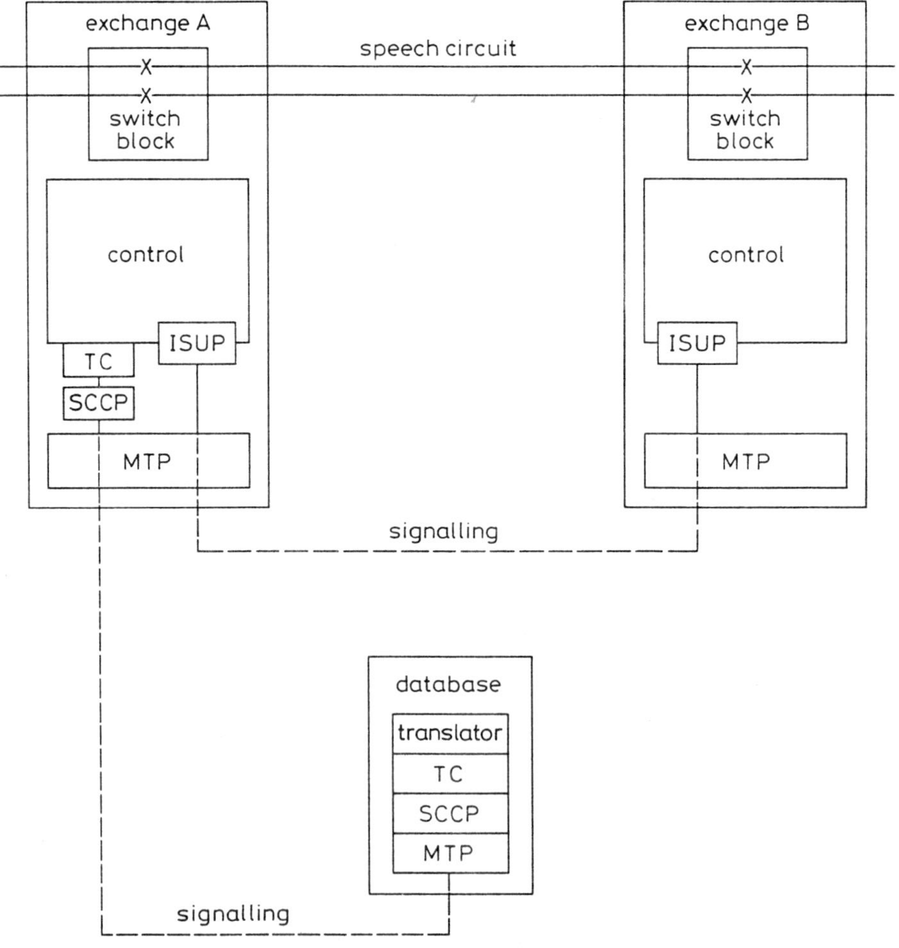

Fig. 4.13   Example of use of elements of architecture

the required routeing translation and returns the information to the control function of Exchange A via TC, SCCP and MTP. With the new information, the control function of Exchange A can now request the ISUP to set up a speech circuit from Exchange A to Exchange B. This is achieved by appropriate communication between the ISUP of Exchange A and the ISUP of Exchange B. Further detail of a similar example is given in Chapter 10.

4.8 Application of the layer structure to DSS1

4.8.1 General

The CCITT Digital Subscriber Signalling System No. 1 (DSS1) is designed for use in an ISDN. Signalling information between customers and the network is carried on a 'D Channel'. The CCITT [8, 9] defines a D channel as being either:

(*a*)   A 16 kbit/s channel used to control two 64 kbit/s channels or
(*b*)   A 64 kbit/s channel used to control multiple 64 kbit/s channels (e.g. thirty 64 kbit/s channels in a 2 Mbit/s transmission system).

The architecture of DSS1 is not complicated by the 4-level structure originally adopted for CCITT No. 7. The international specification of DSS1 [10] is aligned with the OSI model. DSS1 provides the functions of Layers 1 to 3 of the model, compatible with the customer-to-customer context described in Section 4.6. The layers of DSS1 are explained in Sections 4.8.2 to 4.8.4.

DSS1 must handle a variety of circumstances at customers' premises. For example, a local exchange might need to send signalling information to a specific terminal (e.g. a telephone) at a customer's premises. This form of working is termed 'point-to-point' and is illustrated in Fig. 4.14. Whilst there is a common physical connection between the terminals and the local exchange , there is the possibility of separate signalling between each terminal and the local exchange. Thus, the exchange can communicate with either of the terminals separately. Alternatively, a local exchange might need to send signalling information simultaneously to a number of terminals at a customer's premises (e.g. a number of data terminals). This form of working is termed 'broadcast' and the concept is illustrated in Fig. 4.15. Again, there is a common physical connection between the terminals and the local exchange. In this case, there is also a common signalling connection, allowing the exchange to contact both terminals simultaneously. Procedures and formats in DSS1 cover both types of working and more information is given in Chapters 8 and 9.

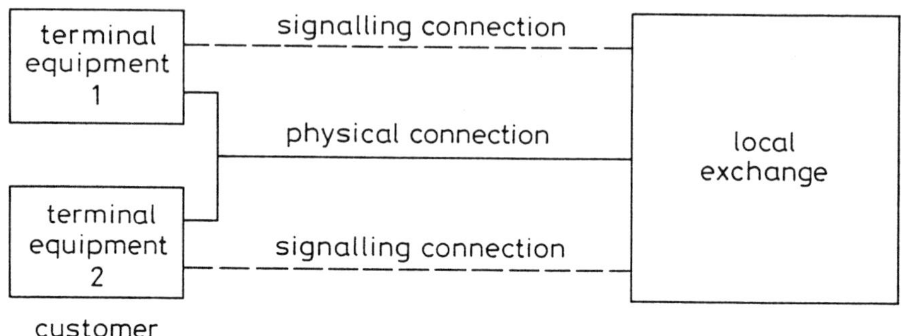

Fig. 4.14   Point-to-point working
Source: CCITT Recommendation Q. 920

DSS1 is also required to handle the transfer of 'packet' data [11, 12]. This is a form of data transfer in which data is divided into blocks called packets. Each packet is routed independently through the network, the blocks being recombined when they have been received at the intended destination. In this way, transmission capacity within the packet network is allocated to a call only when required to transmit a packet. Between successive packets relating to a call, the transmission capacity is used to convey packets for other calls. This differs from circuit-switched calls (e.g. telephony), in which transmission capacity (a traffic circuit) is allocated to a call for the duration of that call. Details of packet data are beyond the scope of this book, but the means by which DSS1 is specified to handle packet data is covered in Chapter 9.

4.8.2 Layer 1

Layer 1 (physical layer) of DSS1 defines the physical, electrical and functional characteristics of the transmission link. The physical layer provides the physical connection for the transmission of bits and allows the transfer of messages generated by Layers 2 and 3. Further information is given in Chapter 8.

4.8.3 Layer 2

When functions within Layer 3 at a customer's premises need to communicate with the network, an association (or connection) is established to allow information transfer to occur. Such associations are termed 'data-link connections'. Layer 2 provides the ability to establish and control one or more data-link connections on a D channel.

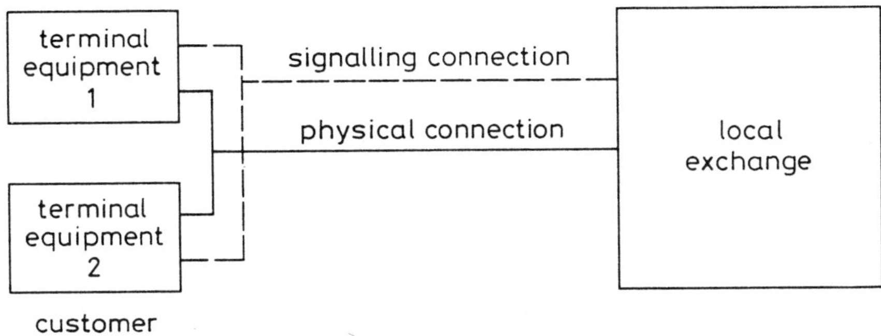

Fig. 4.15   Broadcast working
Source: Recommendation Q. 920

The functions of Layer 2 include sequence control (to maintain the sequential order of messages across a data-link connection) and the detection and correction of errors in messages transmitted on the data link. Information is transmitted in 'frames', which are equivalent to the signal units described for CCITT No. 7 in Chapter 5.

The points at which Layer-2 services are offered to Layer 3 are termed 'service-access points', each point being identified by 'service-access-point identifier (SAPI)'. Many customer terminals can be associated with a SAPI and, to identify a particular terminal, each terminal is allocated a number called a 'terminal end-point identifier (TEI)'. The point-to-point form of working can be implemented by selecting the TEI of the appropriate terminal. The broadcast form of working is implemented by selecting a TEI reserved for the purpose.

Two types of operation are defined for the data-link layer. These are 'acknowledged operation' and 'unacknowledged operation'. In acknowledged operation, each frame transmitted by Layer 2 is numbered. This allows the data-link layer to acknowledge each frame that is received. If errors are detected, or a frame is missing, retransmission of the frame occurs. Procedures, termed 'flow control', also exist in acknowledged operation to limit the number of frames being transmitted. These procedures avoid overloading equipment at the exchange or customer's premises, thus improving the success rate of frame transmission. Acknowledged operation is only applicable to point-to-point working.

In unacknowledged operation, Layer 3 information is transmitted in frames that are not numbered and Layer 2 does not provide an acknowledgement of each frame received.

4.8.4 Layer 3

Layer-3 (network layer) functions are responsible for controlling the establishment and release of calls (both circuit-switched and packet). The functions support the control of basic calls (e.g. basic telephony) and those involving supplementary services. Layer 3 is also responsible for providing transport capabilities additional to those defined in Layer 2. An example of an additional transport capability is the option to re-route signalling messages on an alternative D channel (when provided) in the event of the failure of the normal D channel. It is Layer 3 that generates and interprets messages that are transported by Layer 2. This involves the processing of primitives exchanged with Layer 2, the administration of call references used for call control and ensuring that the services provided by Layer 3 are consistent with the requirements of the customer.

Functions provided by Layer 3 include:

(*a*)   Routeing messages to an appropriate destination (typically a local exchange)

(*b*)   Conveying user-to-user information, with or without the establishment of a circuit-switched connection

(*c*)   Multiplexing multiple calls onto a single data-link connection

(*d*)   Segmenting and reassembling messages to allow their transport by the data-link layer

(*e*)   Detecting errors in the procedures defined for Layer 3, interpreting and reacting to errors detected by the data-link layer and detecting errors in messages provided by the data-link layer operating in the unacknowledged method

(*f*)   Ensuring that messages are delivered to the destination customer in the same order as they are generated by the originating customer.

4.8.5 Primitives

The primitive types used by DSS1 are request, indication, response and confirmation, as described in Section 4.5. Their application conforms to the general description outlined in Section 4.5.9.

In the data-link layer unacknowledged method of operation, Layer 3 requests Layer 2 to transfer information using a primitive called 'unitdata-request'. The 'unitdata-indication' primitive is passed from Layer 2 to Layer 3 to present information received by the data-link layer.

In the data-link layer acknowledged method of operation, the primitive types are qualified by 'establish' and 'release' to indicate the need to set-up and clear-down a Layer 2 connection. The 'data-request' and 'data-indication' primitives are used in a similar manner to the unitdata primitives above.

4.9 Specification of CCS systems

The requirements in Section 4.2 led to the tiered architecture of modern-CCS systems described in Sections 4.3 to 4.8. The specification of CCS systems is based upon the tiers described. In the context of CCS specifications, the customers, exchanges and nodes in the network can be termed 'entities'. CCS systems are defined by specifying:

(*a*)   The procedures used between a tier in one entity and the corresponding tier in another entity. Procedures define the logical sequence of events and flows of messages to provide services.

(*b*)   The format of messages used to enact the procedures between a tier in one entity and the corresponding tier in another entity. The format defines the general structure of messages and specific codings of fields within messages.

(*c*)   The primitives used between adjacent tiers within an entity. By defining the primitives, the interface between adjacent tiers can be held stable even if the functions performed by one tier change.

The application of items (*a*), (b) and (c) to part of a CCS system (e.g. Transaction Capabilities) is termed a 'protocol'. Subsequent chapters describe CCITT No. 7 and DSS1 in terms of the specified procedures, formats and primitives for various tiers. Further details of the use of these techniques are given in Chapter 10.

The adoption of this disciplined approach has enabled the production of comprehensive and high-quality specifications for CCITT No. 7 and DSS1. Many experts throughout the world have contributed views and ideas, resulting in CCS systems that are flexible in meeting the needs of customers. However, despite all these measures, it is still notoriously difficult to produce a specification that can be implemented by manufacturers and network operators and yet be completely unambiguous. It is a stringent test when two exchanges built by one manufacturer are connected together with CCITT No. 7. However, the ambiguous nature of specifications can be masked in these circumstances. The interpretation of a particular clause in the specification is the same for the two exchanges because they are designed by the same people. Thus, even if the interpretation is incorrect, it is still possible that the signalling system performs adequately. The true test is when two exchanges built by two independent manufacturers are connected together. In this case, the interpretation of the specification applied to one exchange is independent of the interpretation applied to the other exchange. If the specification is ambiguous, and different interpretations have been applied, the signalling system will exhibit major abnormalities and may not work at all.

The moral is to learn from practical experience and feed the results of this experience back into the specifications, thus continuously improving the definitions of the CCS systems. This process is taking place and has already highlighted areas that were thought to be clear but, in hindsight, are recognised as needing improvement. Early trials of the message-transfer part (MTP) between the United Kingdom and Belgium [13] illustrate the point. A problem that was discovered very early in the testing programme related to the flags that are used to delimit the CCITT No. 7 messages. One implementation had interpreted the specification as stating that only one flag was allowed between messages. The other implementation interpreted the specification as stating that several flags could be used between messages. These differing interpretations resulted in one implementation not being able to handle multiple flags and the other implementation not being able to handle single flags, thus resulting in an inability to adopt CCITT No. 7 between two implementations. This discovery was, in itself, a major benefit to the specification process because the relevant CCITT recommendation was quickly amended to avoid problems in the future. The problem was also quickly overcome in practice and the field trial was able to continue. It is of interest to note that the majority of the problems encountered during the early field trials were due to design anomalies, rather than specification ambiguities. This reflects the high quality of specification that has been attained.

When it is difficult to reach agreement on how to specify a particular feature in CCS systems, or a network operator has a particular problem that is not common to many others, 'options' can be specified that give alternative methods of implementing the feature. Early field trials reinforced the view that such options must only be specified in exceptional circumstances. Too many options complicate the specification and increase the complexity of interworking different implementations. Even worse, options represent an inability to achieve a true standard that is universally accepted and thus they dilute the benefits of a common specification. Flexibility is a key requirement of CCS systems but diverse ways of achieving flexibility negate the advantages of a common approach. If options are essential, they must be clearly identified as options within the specification and the implications of interworking between different implementations must be fully evaluated.

New customer requirements arise continuously and CCS systems must respond to the challenge of supporting them. Changes to specifications can be minimised by adopting the structured architecture described earlier, but there will always be occasions when new features, or new ways of implementing features, are required. Network operators must respond quickly to customer needs and, to avoid adoption of varying interim systems, the derivation of a standard must be equally fast. In the past, the derivation of standards has taken too long and it has been necessary for network operators to implement non-standard systems in the interim period. The objective must be to increase the speed of agreement on standards to match the needs of customers.

4.10 Chapter summary

Modern CCS systems are specified in a structured manner to ensure that they can flexibly meet customer requirements and exhibit evolutionary potential. The structured approach to specification is known as the architecture of the signalling system.

To allow flexibility and evolution, the architecture must allow one part of the signalling system to be modified without affecting other parts. The signalling system must cater for a range of services and it must be able to handle a multitude of interworking cases. The architecture must also support the speedy production of unambiguous specifications.

The architecture adopted for modern CCS systems is based on dividing signalling functions into a number of tiers. The tiers comprise self-contained groups of functions. A user can be a customer or a node in the network. Specifications are derived by defining (*a*) the primitives used between adjacent tiers within a user, (*b*) the procedures adopted for transferring information between corresponding tiers in different users and (*c*) the format of the information transferred between corresponding tiers in different users.

CCITT No. 7 has been specified using two different types of tier structure. Circuit-related functions are specified using a 4-level structure, whilst non-circuit-related functions are specified using the OSI 7-layer structure. Levels 1 to 3 of CCITT No. 7 form the message-transfer part (MTP), which is responsible for transferring messages from one signalling point to another, even in the event of link and network failures. Level 4 comprises the user parts, which define the meaning of the messages transferred by the MTP and the logical sequence of events for a particular application (e.g. telephony).

Layers 1 to 3 of CCITT No. 7 comprise a combination of the MTP and the signalling-connection-control part (SCCP). The combination of the MTP and SCCP provides a network-layer service to higher layers. Layers 4 to 7 of CCITT No. 7 comprise transaction capabilities (TC). TC is a protocol designed to carry non-circuit-related data between nodes in the network.

DSS1 is specified only in terms of Layers 1 to 3 of the OSI 7-layer structure. Layer 1 covers the physical and electrical conditions of the transmission link. Layer 2 establishes data-link connections to convey signalling information. Layer-2 operation can be in an acknowledged or unacknowledged form. To cater for a variety of terminal arrangements at customers' premises, Layer 2 allows for a point-to-point form of working (in which communication with a particular terminal is required) and a broadcast form of working (in which signalling information is transmitted to all terminals). Layer 3 of DSS1 defines the meaning of the messages transferred by Layer 2 and the logical sequence of events that can occur during call set-up and release.

The international specification process makes use of experts throughout the world to produce high quality standards. It is very difficult to achieve unambiguous specifications and it is essential that feedback from practical implementations is used to achieve continuous improvement. Early trials of CCITT No.7 were used to this effect. It is essential that the number of options allowed in specifications is kept to a minimum. It is also essential that international specifications are produced faster to obviate the need for interim implementations to meet evolving customer requirements.

4.11 References

1 MANTERFIELD, R J: 'Specification and evolution of CCITT No. 7', Proceedings International Switching Symposium, Phoenix, 1987
2 CCITT Recommendation Q.701: 'Functional description of the MTP of Signalling System No. 7' (ITU, Geneva)
3 JENKINS, P A and KNIGHTSON, K G: 'Open systems interconnection - an introductory guide', *British Telecommunications Engineering Journal*, 1984, **3**, pp. 86–91
4 CCITT Recommendation X.200: 'Reference model of open systems interconnection for CCITT applications' (ITU, Geneva)
5 International Standards Organisation : IS7498, 'Information processing systems - Open systems interconnection - Basic reference model'.
6 CCITT Recommendation Q.700: 'Introduction to CCITT Signalling System No.7' (ITU, Geneva)
7 CCITT Recommendation X.229: 'Remote operations protocol' (ITU, Geneva)
8 CCITT Recommendation I.430: 'Basic user-network interface Layer 1 Specification' (ITU, Geneva)
9 CCITT Recommendation I.431: 'Primary rate user-network Layer 1 specification' (ITU, Geneva.)
10 CCITT Recommendations Q.920, Q.921, Q.930 and Q.931: 'Digital Subscriber Signalling System No. 1' (ITU, Geneva)
11 CCITT Recommendation X.25: 'Interface between data terminal equipment and data circuit terminating equipment for terminals operating in the packet mode' (ITU, Geneva)
12 CCITT Recommendation X.75: 'Packet switched signalling system between public networks providing data transmission service' (ITU, Geneva)
13 BEKAERT, MANTERFIELD, THOMAS and HAERENS: 'Testing results from the implementation of the CCITT No. 7 Signalling System', International Switching Symposium, 1984

Chapter 5

CCITT No. 7 transfer mechanisms

5.1 Introduction

The transfer of signalling information between signalling points is achieved by the message-transfer part (MTP) and the signalling-connection control part (SCCP). An explanation of the terminology is given in Chapter 4. The MTP and SCCP do not understand the meaning of the messages being transferred, but it is their job to deliver the information in an uncorrupted form from one signalling point to another. For circuit-related applications (e.g. telephony), the MTP provides an adequate transfer mechanism. For non-circuit-related applications (e.g. general data transfer), a combination of the MTP and SCCP is required.

The architecture of CCITT No.7 is described in Chapter 4. The MTP comprises Levels 1 to 3 of the 4-level structure. The combination of the MTP and SCCP lies within Layers 1 to 3 of the OSI 7-layer model. Sections 5.2 to 5.5 describe the functions of the MTP and Section 5.6 covers the SCCP.

5.2 Message-transfer part

The message-transfer part (MTP) [1,2] is responsible for the transfer of signalling information from one signalling point to another in circuit-related applications. The aim is to deliver the information in messages without loss, without duplication, free of errors and in a pre-arranged sequence. The specification of the MTP is structured to allow flexible implementations that can be optimised in practical networks. It is an intelligent transfer mechanism that can reconfigure and control signalling traffic to overcome failures in the network. The user parts, by defining the meaning of messages and high-level procedures, add the overall intelligence needed to transform the transfer mechanism provided by the MTP into an effective circuit-related communications medium.

The MTP is responsible for a very high level of reliability for successful conveyance of messages. To achieve the high targets set by CCITT[3], error-detection and error-correction techniques are employed, as well as actions to control the signalling network. Some significant performance targets are:

(*a*)   Undetected errors; less than one in 10^{10}
(*b*)   Loss of messages; less than one in 10^7
(*c*)   Out-of-sequence delivery to higher levels; less than one in 10^{10}.

The structured nature of the MTP allows it to be described in terms of the three levels shown in Fig. 5.1.

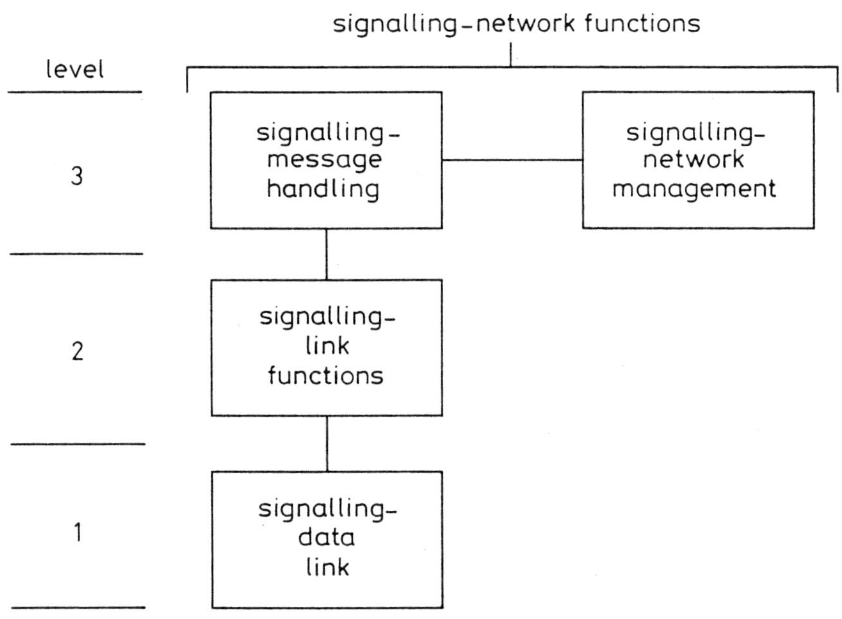

Fig. 5.1   Levels of the message – transfer part

5.3 MTP Level 1: signalling-data link

Level 1 defines the physical, electrical and functional characteristics of the transmission path for signalling. It is normally a 64 kbit/s path within a PCM system, but other forms of transmission (including analogue) can be used. The provision of a Level 1, defining the interface to the transmission medium, means that the higher levels (Levels 2–4) are independent of the type of transmission medium used.

5.4.1 General

Level 2 defines the functions and procedures for the transfer of signalling messages over a signalling-data link between two directly-connected signalling points. These terms are described in Chapter 4. The combination of Levels 1 and 2 provides a signalling link for reliable transfer of signalling messages. The Level-2 functions provide a framework for the information transferred over each link and perform error-detection and error-correction procedures.

Signalling information transferred between signalling points is divided into messages termed 'signal units'. These signal units vary in length according to the type of information being transferred.

There are three types of signal unit:

(a)   The message signal unit (MSU), which is used for transferring signalling information supplied by a user part or SCCP.

(b)   The link-status signal unit (LSSU), which is used to indicate and monitor the status of the signalling link.

(c)   The fill-in signal unit (FISU), which is used when there is no signalling traffic to maintain link alignment.

The three types of signal unit have a very similar format. The type of signal unit is identified by a length indicator (LI) as follows:

LI = 0, fill-in signal unit
LI = 1 or 2, link-status signal unit
LI greater than 2, message signal unit.

The most comprehensive signal unit is the message signal unit (MSU) and its format is shown in Fig. 5.2. It is shown that the MSU is divided into a number of fields, with a specified number of bits allocated to each field. Hence, the format of the MSU defines each of the fields within the message and allocates a meaning to each bit within the message. The exception to this is the signalling-information field, which is defined by Level-3 functions. The principles of Level-2 functions can be explained by considering the fields within the MSU. These are described in Sections 5.4.2 to 5.4.7. Details of the error-correction mechanisms used in the MTP are given in Sections 5.4.8 to 5.4.10.

5.4.2 Flag

The flag acts as a signal-unit delimiter, the beginning and end of each signal unit being indicated by a unique 8-bit pattern. In typical implementations, the end flag of one MSU acts as the beginning flag of the next MSU. The

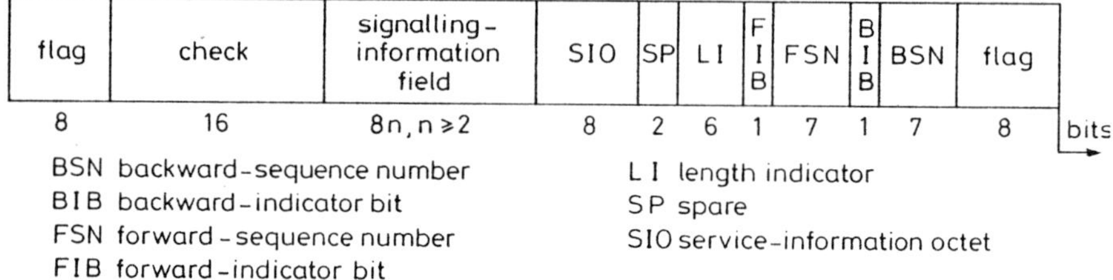

Fig 5.2   Format of message-signal unit
Source: CCITT Recommendation Q. 703

pattern is coded 01111110. To avoid the flag being imitated by another part of the signal unit, the exchange transmitting the MSU inserts a zero after every sequence of five consecutive ones occurring in any position in the MSU apart from the flag. This zero is deleted at the receiving end of the signalling link after the flags have been detected.

5.4.3 Sequence numbers
The backward-sequence number, backward-indicator bit, forward-sequence number and forward-indicator bit are used in the error-correction mechanism described in Section 5.4.8.

5.4.4 Length indicator
The length indicator (LI) gives the length of the signal unit. A LI value of greater than 2 indicates that the signal unit is a MSU.

5.4.5 Service-information octet
The service-information octet (SIO) defines the user part, or equivalent, that is appropriate to the message. For example, the SIO can indicate that the message is relevant to the telephone, data or ISDN user part or to the SCCP.

5.4.6 Signalling-information field
The signalling-information field (SIF) may consist of up to 272 octets, the formats and codes being defined by the user part. Early designs of CCITT No. 7 may use a maximum signalling-information field of 62 octets, in line with early specifications for the MTP. The SIF contains the information that needs to be transferred between the user parts of two signalling points. Thus, the MTP is not aware of the contents of the SIF, except for the routeing label, which is information that is used for routeing messages in the

signalling network (see Section 5.5.2). Apart from such routeing information, the MTP merely transfers the information in the SIF from Level 4 of
one exchange to Level 4 of another exchange.

5.4.7 Error detection

Error detection is performed by means of 16 check bits provided at the end
of each signal unit. The check bits are derived by the exchange sending the
signal unit. The check bits are generated by applying a complex polynomial
to the information in the signal unit. The polynomial used is:

$x^{16} + x^{12} + x^5 + 1$

This polynomial is chosen to optimise the detection of bursts of errors
during transmission. The check bits are formed by:

(a)   The remainder of $x^k (x^{15} + x^{14} + x^{13} + x^{12} + ... + x^2 + x + 1)$ divided by
      the polynomial $x^{16} + x^{12} + x^5 + 1$, where k is the number of bits in the
      signal unit, and

(b)   The remainder after multiplication by x^{16} and then division by the
      polynomial $x^{16} + x^{12} + x^5 + 1$, of the content of the signal unit.

The check bits transmitted are the ones complement of the resulting 16-bit
field, i.e. the ones are changed to zeros and vice versa. This change is
performed to minimise the risk of faulty operation of the receiving-
exchange equipment.

The check bits are analysed at the receiving exchange according to a
corresponding algorithm. If consistency is not found, then an error has been
detected and the message is discarded. Discarding a MSU in this way invokes
the error-correction mechanism outlined in Section 5.4.8.

5.4.8 Error correction

There are two error-correction methods defined for CCITT No. 7 [4]. The
'basic' method is appropriate for links with one-way propagation delays of
less than 15 ms and the 'preventive-cyclic-retransmission' method is appropriate for links with one-way propagation delays of greater than 15 ms.

Using these techniques, messages that have been corrupted (e.g. due to
error bursts on the transmission medium) are retransmitted in sequence.
Level 3 is unaware of any problems if the error- correction mechanism is
successful. In this case, the messages are delivered to the user part without
loss or duplication. If persistent faults occur, Level 3 is informed so that
management action can be taken. An example of such an action is re-
routeing messages via different signalling links. The basic method of error
correction is described in Section 5.4.9 and the preventive-cyclic-retransmission method is described in Section 5.4.10.

5.4.9 Basic method of error correction

The basic method of error correction is a 'non-compelled, positive/negative acknowledgement, retransmission error-correction system'. Non-compelled means that messages are sent once only, unless they are corrupted during transfer. Positive/negative acknowledgement means that each message is acknowledged as being received, with an indicator to explain whether or not the message is corrupted. Error correction is by means of retransmission. The functions involved in the error-correction mechanism are shown in Fig. 5.3.

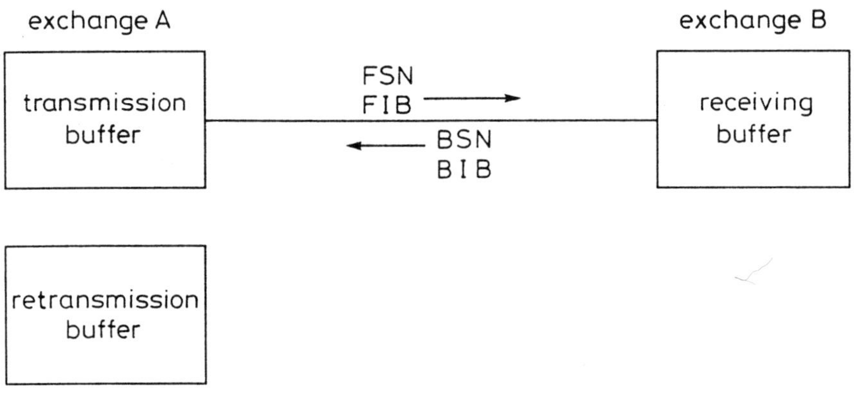

Fig. 5.3   Error correction functions

When Level 4 of Exchange A needs to send signalling information to Level 4 of Exchange B, the relevant information is passed (via Level 3) to Level 2 of Exchange A. Level 2 of Exchange A is provided with a transmission buffer and a retransmission buffer. The transmission buffer is used to store the MSU before sending it on the signalling link. Hence, the transmission buffer acts as a store until signalling-link capacity is available to send the MSU. The retransmission buffer keeps a copy of the MSU in case of corruption during the transfer of the MSU to Exchange B.

Each MSU contains a forward-sequence number (FSN), a forward-indicator bit (FIB), a backward-sequence number (BSN) and a backward-indicator bit (BIB). When the signalling link is acting normally, the FIB is set to a particular value (e.g. zero) and the BIB is set to the same value (zero). When a MSU is received by Level 2 at Exchange A, it is entered into the transmission buffer. The transmission buffer acts as a queue, working on the principle that the first MSU received is the first to be transmitted. When the signalling link is free, and it is the turn of the MSU in the example to be sent, the MSU is allocated a FSN based on the last FSN plus 1 (modulo 128). The

MSU is then transmitted to Exchange B. A copy is also entered into the retransmission buffer.

At the receiving buffer at Exchange B, the FSN is compared with the expected value (last FSN + 1). If the FSN is the expected value, the MSU is passed to Level 3 for processing. The FSN value is copied into the BSN field and the BIB is left unaltered. The values of BSN and BIB indicate a positive acknowledgement to Exchange A. Upon receipt of the correct BSN and BIB values at Exchange A, the MSU is deleted from the retransmission buffer.

If the comparison of FSNs at Exchange B shows a discrepancy (e.g. due to the error-detection mechanism discarding a MSU), the value of the BIB is inverted, (i.e. changed to value 1), thus denoting a negative acknowledgement to Exchange A. In this case, the BSN is made the value of the last correctly-received FSN. Upon receipt of a negative acknowledgement at Exchange A, the transmission of signal units is interrupted and the MSUs in the retransmission buffer are retransmitted in the original order. The value of the FIB is inverted (i.e. changed to value 1) so that the FIB and BIB are once again the same value.

5.4.10 Preventive-cyclic-retransmission method of error correction

This method is a 'positive acknowledgement, cyclic retransmission, forward error-correction system'. This means that negative acknowledgements are not used, the mechanism relying on the absence of a positive acknowledgement to indicate corruption of the message. Forward error correction is achieved by a programmed cycle of retransmission of unacknowledged MSUs. Each signal unit contains a FSN and BSN (as for the basic method), but the FIB and BIB are not examined and are permanently set to value 1. When no new MSUs are available for transmission, a cyclic retransmission of all MSUs remaining in the retransmission buffer is commenced. The original FSNs are maintained during the retransmission. If a new MSU arrives, the cyclic retransmission is stopped and the new MSU is transferred with a FSN of the last new MSU plus 1 (modulo 128). If no further new MSUs are received, the cyclic retransmission is recommenced.

An uncorrupted MSU is positively acknowledged by means of the receipt at Exchange A of a BSN equal to the allocated FSN. After positive acknowledgement, the relevant MSU is discarded from the retransmission buffer and is no longer available for retransmission.

One disadvantage of this system is that the sending transmission and retransmission buffers could become overloaded. To prevent message loss, a procedure called forced retransmission is adopted. The number of MSUs, and the number of octets in the retransmission buffer, are monitored continuously. If either parameter reaches a pre-set threshold, new MSUs are not accepted and priority is given to cyclic retransmission of MSUs in the retransmission buffer. The retransmission cycle is continued until the two activating parameters fall below the specified thresholds.

5.5 MTP Level 3: signalling-network functions

5.5.1 General

Chapter 1 describes the drive for CCS systems to be able to provide unimpeded communication between exchange processors. For maximum flexibility, separate signalling networks are evolving. It is Level 3 that provides a major step in the ability for CCITT No. 7 to achieve a cohesive signalling network[5].

Level 3 functions are responsible for the reliable transfer of signalling information from one exchange to another, even in the case of Level-1 or Level-2 failures. Level 3 is responsible for those functions that are appropriate to a number of signalling links, including controlling signalling links and routeing messages through the network. The functions are divided into 'signalling-message handling', which covers the routeing of messages through the signalling network (described in Section 5.5.2) and 'signalling-network management', which covers the control of the signalling network itself. The signalling-network management functions can be divided into:

(*a*)   *Signalling-traffic management*, covering the reconfiguration of signalling traffic in response to changes in network status. This is described in Section 5.5.3.

(*b*)   *Signalling-link management*, which controls the signalling links. This is described in Section 5.5.4.

(*c*)   *Signalling-route management*, which covers the distribution of information on the signalling network status. This is described in Section 5.5.5.

5.5.2 Level 3: signalling-message-handling function.

This function defines how signal units are routed through the signalling network. Each signalling point within a signalling network is identified by a 'point code', i.e. a 14-bit code or address that is unique within the signalling network. The destination-point code (DPC) identifies the destination signalling point of the message and the originating-point code (OPC) identifies the originating signalling point.

When signalling information is received by the message-handling function from Level 4, the information includes a routeing label. The structure of the routeing label is shown in Fig. 5.4. By analysing the DPC, the message-handling function can determine to which exchange the signal unit should be sent. Hence, this analysis allows the choice of an appropriate signalling link. If two or more signalling links to the required destination exist, the message-handling function performs a load-sharing activity over the links. In this case, the signalling link selection (SLS) field is used to identify the chosen signalling link and hence perform the load sharing. The SLS consists of the 4 bits following the OPC.

When a signal unit is received by the message-handling function from Level 2, the DPC is analysed to determine whether or not the signal unit is destined for another signalling point. If the message is destined for the receiving signalling point, the message is delivered to the appropriate user part. This is determined by analysing the service information octet (Fig. 5.2). If the message is destined for another signalling point, the analysis of the DPC indicates how to route the signal unit. In this case, the signalling point performing the analysis of the DPC is acting as a signal-transfer point. One important aspect is that the message is re-routed without Level 4 needing to know. Thus, the signal-transfer point working avoids a significant processing overhead on each message.

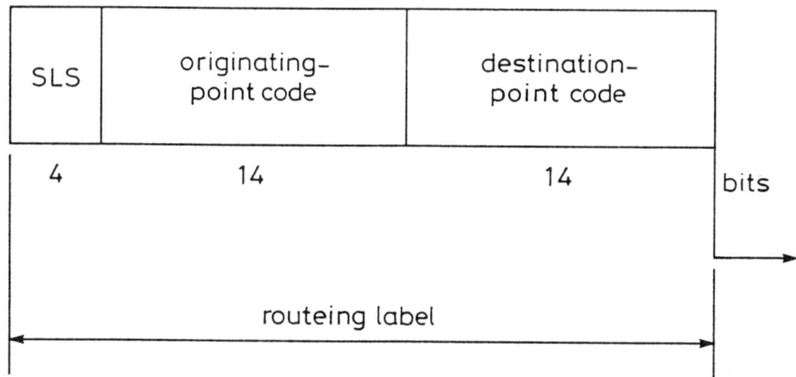

Fig. 5.4   Structure of routeing label
SLS: signalling-link selection

5.5.3 Level 3: signalling-traffic-management function

The signalling-traffic-management function of Level 3 provides the procedures required to maintain a flow of signalling traffic in the event of disruptions in the signalling network. Such disruptions include the failure of signalling links and the failure of signal-transfer points.

The signalling-traffic-management function is used to divert signalling traffic. In addition, in the case of congestion at a signalling point, the function is used to reduce the traffic to that signalling point. To achieve these results, several procedure are defined. Examples of significant procedures are 'changeover', 'changeback', 'management inhibit' and 'signalling-traffic flow control'.

The changeover procedure ensures that signalling traffic is diverted from a signalling link, to an alternative signalling link, as quickly as possible. A typical example of when the changeover procedure is initiated is when a signalling link fails. The changeover must be implemented without message loss, duplication or mis-sequencing. This is achieved by implementing measures to ensure that messages in the retransmission buffer of the unavailable signalling link are transferred to the alternative link. If an alternative signalling link is not available, the destination is regarded as inaccessible and the user part is informed accordingly.

The changeback procedure reverts the routeing of signalling messages back to the conditions prevailing before changeover. Measures are included to control the message sequence to ensure that messages are not lost or mis-sequenced. Changeback is initiated when the conditions causing changeover have been rectified, e.g. when a signalling link is restored.

The management-inhibit procedure is used to facilitate maintenance or testing. The procedure does not cause a change of status at Level 2, thus leaving the link available to send maintenance and test messages. If failure conditions occur in the signalling network, resulting in the need to use the inhibited signalling links, the inhibiting procedure can be over-ridden and the signalling links put back into service.

Signalling-traffic flow control is used to limit signalling traffic at its source when the signalling network is unable to transfer all signalling traffic offered by user parts.

5.5.4 Level 3 : signalling-link-management function

This function is used to control the signalling links connected to a particular exchange. The function provides a means of establishing and maintaining the capability of signalling links. In the event of signalling-link failures, the signalling-link- management function controls the actions aimed at restoring full capability.

The basic procedures defined for signalling-link management are 'signalling-link activation' and 'signalling-link restoration'. Signalling-link activation is the process of making a signalling link ready to carry traffic. It involves establishing alignment on the signalling link and performing a test to ensure correct functioning. Signalling-link deactivation is the process of taking a link out of service. Signalling-link restoration is similar to signalling-link activation, but it applies to re-introduction of service on a signalling link after failure.

5.5.5 Level 3 : signalling-route-management function

A signalling 'route' is a collection of signalling links connecting two signalling points. It may be a direct route, in which signalling links are directly connected, or an indirect route, in which signalling is via a signal-transfer point.

The signalling-route-management function is used to distribute information about the signalling network status in order to block (i.e. prevent access to), or unblock, signalling routes. Significant procedures are 'transfer prohibited', 'transfer allowed' and 'signalling-route-set test'. The transfer-prohibited procedure is initiated by a signal-transfer point. The aim is to notify one or more adjacent signalling points that messages should not be routed via the signal-transfer point to a particular destination. The transfer-allowed procedure is used to remove the prohibited status.

The signalling-route-set-test procedure is initiated by a signalling point to test whether or not signalling traffic towards a particular destination can be routed via a signal-transfer point. The signalling-route-set-test message contains the current route status of the destination, as understood by the sending-signalling point. On receipt of the message, the signal-transfer point compares the status of the destination specified in the received message with the actual status. If they are the same, no action is taken. If they are different, the result is returned to the signalling point.

5.5.6 Signalling network

The functions described above illustrate the ability of the MTP to transfer messages between signalling points by providing a comprehensive signalling network. The use of signal-transfer points enhances the flexibility of the MTP. Even during failure conditions within the traffic network or the signalling network, the MTP can reconfigure message routeing to ensure the delivery of messages to the correct destination without corruption or duplication.

5.6 Signalling-connection-control part (SCCP)

5.6.1 General

The MTP is a comprehensive transfer mechanism that was specified before the open-systems-interconnection (OSI) 7-layer model was derived. To be flexible in future network environments, it is important that CCITT No. 7 fits within the OSI 7-layer model approach for non-circuit-related applications, as explained in Chapter 4.

The MTP adequately covers Layers 1 and 2 of the OSI model, but a number of functions need to be added to the MTP in order to provide adequate Layer 3 functions. Without the additional functions, the MTP does not offer an OSI-network service. The functions additional to the MTP that have been specified to provide a Layer 3 service are collectively called the 'signalling-connection control part (SCCP)'. The combination of the MTP and the SCCP is called the 'network-service part (NSP)'.

The objective of the SCCP is to allow data transfer between two nodes, even when physical speech paths are not involved. Hence, the SCCP allows the non-circuit-related transfer of data across the network. The term 'node' indicates an exchange or any other network element (e.g. a database) that is capable of non-circuit-related signalling.

The SCCP can be described in terms of its functional structure [6], formatting principles [7] and procedures [8], described in Sections 5.6.2 to 5.6.7.

5.6.2 Functional structure

Chapter 4 describes how the SCCP fits into the overall architecture of CCITT No.7 and Fig. 4.9 shows that the SCCP provides the interface between Layers 3 and 4 of the OSI model. Communication between the SCCP and Layer 4 is by the use of primitives. Fig. 5.5 illustrates the primitives associated with the SCCP/ Layer 4 interface. In the context of the OSI 7-layer model, the SCCP is deemed to offer services to the higher layers. The services offered by the SCCP can be categorised into 'connection-oriented' services and 'connectionless' services.

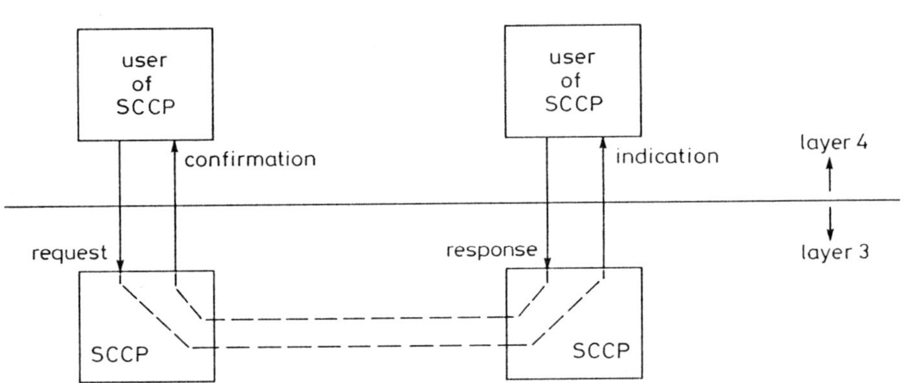

Fig. 5.5   SCCP/user primitives

In connection-oriented services, a relationship is established between two communicating nodes before data transfer begins. The relationship is termed a 'connection'. It is established by exchanging local-reference numbers i.e. numbers allocated by each node to identify to which transaction a message refers. In this case, any data that are transmitted between the nodes include the local-reference numbers and the data are thus associated with the connection. Hence, certain quality of service criteria can be implemented, e.g. in one type of connection-oriented service, it is possible to guarantee to deliver messages in the same order as they are transmitted.

This is possible because the establishment of a connection means that the messages can be correlated by the SCCP.

In connectionless services, the SCCP provides the ability to transfer data via the signalling network without the set-up of a signalling connection. The data are included in a 'unitdata' message transferred between appropriate SCCPs. In this case, it is not possible to guarantee that two messages sent by one node in a particular order will always be received by another node in the same order. This is because they might be routed differently through the signalling network, especially under fault conditions, and there is not a connection that allows the two messages to be correlated by the SCCP.

5.6.3 Formatting for the SCCP

SCCP messages are carried in the signalling-information field (SIF) of message signal units (MSUs), as described in Section 5.4.6. For a MSU carrying a SCCP message, the format of the SIF consists of the routeing label, the message type and parameters. The structure of the SIF for SCCP messages is shown in Fig. 5.6. The routeing label is the standard version described in Section 5.5.2.

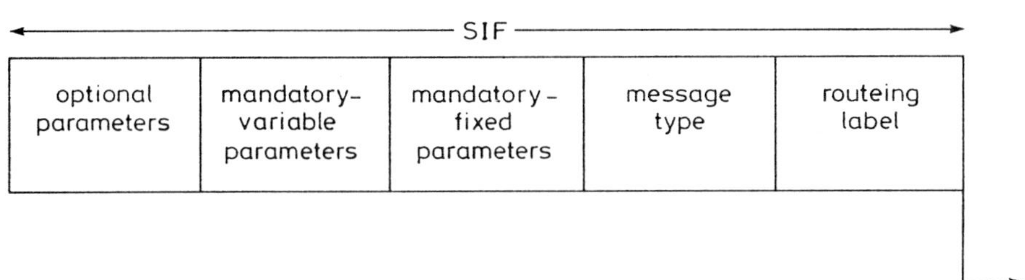

Fig. 5.6   Format of SCCP messages

The message type consists of one octet and it is mandatory for all SCCP messages. The message type uniquely defines the function of the SCCP message. Examples of message types for connection-oriented services include:

(a)   *Connection-Request (CR)*: This message type indicates a request by one node to set up a connection, with given characteristics, with another node.

(b)   *Connection-Confirm (CC)*: This confirms that a connection has been established between two nodes in response to a CR message.

(c)   *Released (RLSD)*: This indicates that the signalling connection is being released by either node.

(d)   *Release Complete (RLC)*: This confirms that the release process is complete.

(e)   *Data Form (DT)*: This message type is used to transfer data transparently between the two nodes, once a connection has been established.

An example of a message type for connectionless services is Unitdata (UDT), which is used to transfer data without establishing a connection between two nodes.

Each message contains a number of parameters that complement the information contained in the message type. In general, each parameter consists of a name, a length indicator and a data field, as shown in Fig. 5.7. The name uniquely identifies the parameter and is coded as a single octet. The length indicator specifies the length of the parameter and the data field contains supporting information. However, not all of these fields are included in all parameters. Parameters can be 'fixed-mandatory', 'variable-mandatory' or 'optional'.

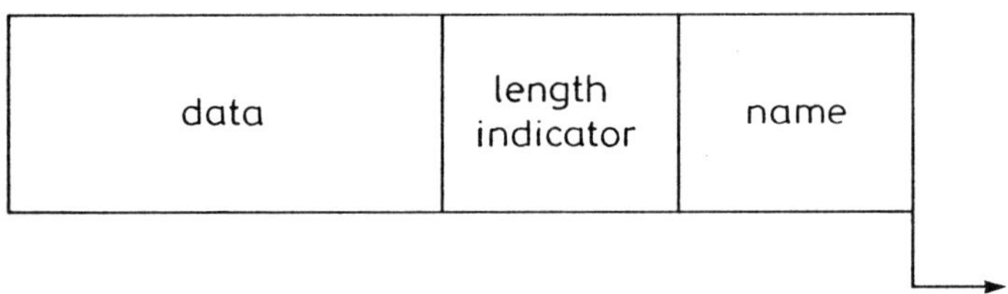

Fig. 5.7   General format of a parameter

Fixed-mandatory parameters must be present in a given message type and they are of fixed length. The position, length and order of fixed mandatory parameters is uniquely defined by the message type, so the parameter names and length indicators are not included in the message. Variable-mandatory parameters must be present in a given message type, but they are of variable length. The name of the parameter is implicit in the message type and hence the parameter name is not included in the message. Optional parameters may or may not occur in any particular message type. Each optional parameter includes the name (one octet) and a length indicator (one octet) before the data field carrying the parameter contents.

To illustrate the principles of SCCP formatting, consider a connection-request message that is used to establish a connection when using an SCCP connection-oriented service. An example of such a message is shown in Fig. 5.8. The message type indicates that the message is a connection request.

The message type is followed by four parameters. The first parameter is a fixed-mandatory parameter called 'source-local reference', which indicates the reference number that the originating SCCP has allocated to identify messages relevant to a particular transaction. The second parameter is also fixed mandatory and it is called 'protocol class', indicating the type of service requested (see Section 5.6.5 for more detail). The third parameter is variable mandatory and it is termed 'called-party address', indicating the identity of the SCCP to which the message is being sent. This parameter includes a length indicator to show the number of address digits included in the data field of the parameter. The fourth parameter is optional and is called

parameter 4 O	parameter 3 V	parameter 2 F	parameter 1 F	message type
calling- party address	called- party address	protocol class	source- local reference	connection request

F – fixed mandatory
V – variable mandatory
O – optional

Fig. 5.8   Example of connection – request message structure

'calling-party address', indicating the identity of the SCCP sending the message. This parameter includes a length indicator and a name.

5.6.4 SCCP procedures

Four types of service, called 'protocol classes', are defined for the SCCP. Two of the classes are termed 'connection-oriented', because they involve the setting-up and clearing-down of signalling connections. In connection-oriented protocol classes, a signalling connection is established, data are transferred and the signalling connection is released upon completion of the data transfer. The data are transferred in blocks called 'network- service data units' (NSDUs), which can be up to 256 octets in length. For longer data streams, the data are split (segmented) into blocks of 256 octets at the sending node so that each block can be transmitted separately. The blocks are then put back together (reassembled) at the receiving node.

The two other classes are termed 'connectionless', because they do not involve the establishment and release of connections. In connectionless protocol classes, the data transfer is inherent within the messages sent across the network. Hence, connectionless protocols do not exhibit set-up and release phases. The maximum length of data is 256 octets, because the connectionless protocols do not perform segmentation and reassembly.

The four protocol classes are outlined in Section 5.6.5 below. Procedures for connection-oriented services are described in Section 5.6.6 and procedures for connectionless services are described in Section 5.6.7.

5.6.5 Protocol classes

Class 0 is a connectionless service. In class 0, each NSDU is transported from the sending SCCP to the receiving SCCP in an independent manner using the MTP. Thus, NSDUs at the receiving node might not be in the same sequence as they were sent. This restriction applies under normal operating conditions, as well as under fault conditions.

Class 1 is also a connectionless service. Class 1 is similar to Class 0, but a limited sequencing mechanism is included. This allows the sending node to request that the NSDUs be delivered to the receiving node in the same sequence as they are sent. The sequencing is performed by the MTP in response to the SCCP selecting a consistent signalling link selection (SLS) field. This technique works under normal operating circumstances; however, under fault conditions in the network, the lack of a connection can still result in mis-sequenced messages.

Classes 2 and 3 are connection-oriented services. In Class 2, NSDUs may be transferred in both directions during the data transfer phase of the connection. Class 3 complements Class 2 by the inclusion of a service guaranteeing that messages will be received in the same order as they were sent, even under fault conditions.

5.6.6 Connection-oriented procedures

An example of the message sequence for connection-oriented services is shown in Fig. 5.9. In this example, higher-layer functions in Node 1 need to communicate with corresponding functions in Node 2. SCCP1 receives a request from a higher-layer function within Node 1 to establish a connection with SCCP2. SCCP1 analyses the called-party address (i.e. the address of SCCP2) and the result of the analysis is that the connection should be established over an appropriate signalling link, via the MTP, to Node 2. A connection-request (CR) message is sent to SCCP2 via the MTP. Upon receipt of the CR message at Node 2, the MTP delivers the CR message to SCCP2. Analysing the called-party address, SCCP2 recognises that the CR message has reached its intended destination and that a connection needs to be established with SCCP1. A connection-confirm (CC) message is returned to SCCP1.

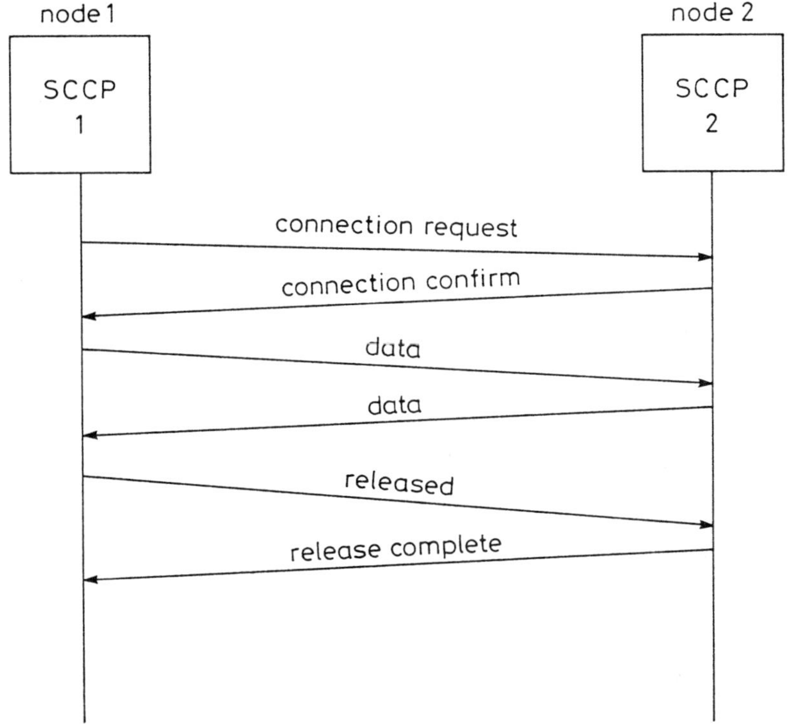

Fig. 5.9   Example of message sequence: connection oriented

When the CR and CC messages have been exchanged, the signalling connection has been established and data transfer can occur. When data transfer is complete, either SCCP1 or SCCP2 can initiate a release procedure by sending a released (RLSD) message. Receipt of the RLSD message by a node is confirmed by returning a release complete (RLC) message.

During connection establishment, source and destination local references are assigned to the connection. The source-local reference is chosen by SCCP1 (from a pool of numbers) and the destination-local reference is chosen by SCCP2. The combination of these local references then acts as the reference number to identify uniquely the SCCP connection. The local references are mandatory fields in SCCP messages. Upon release of the connection, the local references are returned to a common pool at each node and can then be used again for another connection.

The class of protocol can be negotiated during connection set-up. The originating higher-layer function chooses a preferred protocol class and this is included in the CR message sent by SCCP1. The protocol class can be made less restrictive (e.g. move from Class 3 to Class 2) by SCCP2 by marking a field in the CC message to state that only Class 2 is being offered by Node 2. This could be necessary if, for example, Class 3 is unavailable at Node 2.

If Nodes 1 and 2 do not have a direct-signalling relation, it is necessary to involve a third node in the establishment of the connection, as illustrated in Fig. 5.10. In this case, SCCP1 analyses the called-party address and, recognising that there is not a direct-signalling link with SCCP2, sends the CR message to an intermediate SCCP (SCCP3). Upon receipt of the CR message, SCCP3 analyses the called-party address and recognises that the message is intended for SCCP2. Having a direct-signalling link with SCCP2, SCCP3 transmits the CR message to SCCP2. SCCP2 returns the CC message to SCCP1 via SCCP3. In this context, SCCP3 is known as a 'relay point' because of its role in relaying messages between SCCPs 1 and 2.

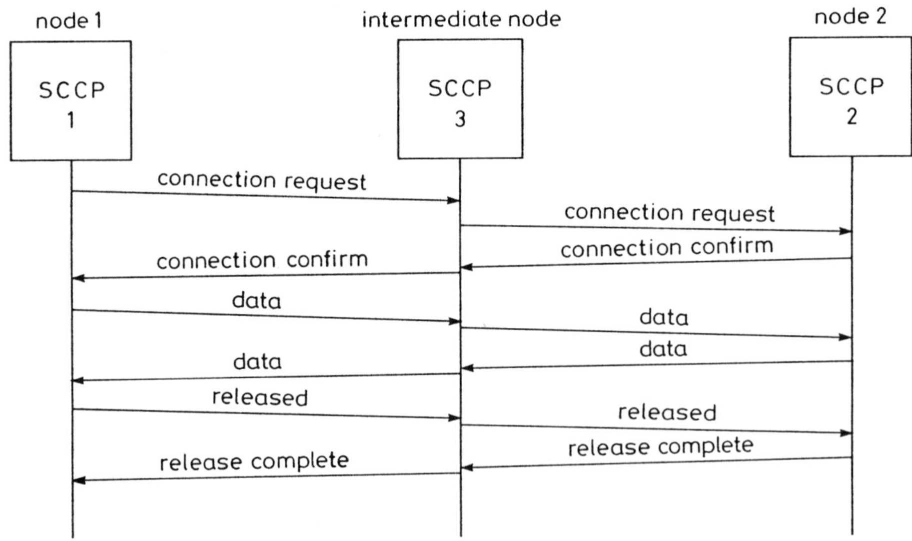

Fig 5.10   Example of message sequence: connection oriented with intermediate node

5.6.7 Connectionless procedures

An example of the message sequence for connectionless services is shown in

Fig. 5.11. In this example, higher-layer functions in Node 1 need to send data to corresponding functions in Node 2 without establishing a connection. SCCP1 receives a request from a higher-layer function within Node 1 to send a NSDU by protocol Class 0 or 1. SCCP1 analyses the called-party address and determines that a message needs to be sent to Node 2. The NSDU is included in a 'unitdata' message transmitted to Node 2. Upon receipt of the unitdata message at SCCP2, an analysis of the called-party address determines that the message has reached its intended destination and the data are delivered to the appropriate higher-layer function.

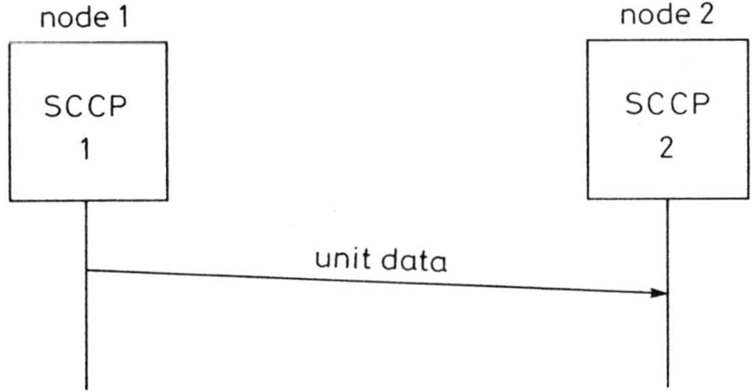

Fig. 5.11   Example of message sequence: connectionless

If intermediate SCCPs are required to route the unitdata message, this is recognised by analysing the called-party address. The result of the analysis indicates the node to which the message should be routed. If the unitdata message is delivered to a node that cannot route the message further, procedures exist either to discard the message or to return the message to the originating node.

5.7 Evolution

The SCCP gives CCITT No.7 the capability of offering an OSI Layer 3/4 interface. Hence, CCITT No. 7 can offer a network service to a range of higher-layer functions. In the short term, this allows CCITT No. 7 the tremendous benefit of meeting the needs of non-circuit-related applications. The specification of SCCP has been derived before widespread implementations have been achieved. Thus, the implementation of standard versions of SCCP throughout the world will help to reduce costs, avoiding the development costs of many different solutions. In the longer term, the benefits could be even greater. By providing a network service, the combination of the MTP and SCCP can be used to provide a transfer capability for a range of higher-layer protocols conforming to the OSI 7-layer model, irrespective of whether or not the higher-layer protocols are specified as part of CCITT No. 7. The possibility will exist to mix and match the various protocols available to network operators, providing great flexibility in implementation and allowing technical solutions to be optimised to meet specific environments. The prospects are fascinating.

5.8 Chapter summary

The message-transfer part (MTP) comprises Levels 1 to 3 of the 4-level structure adopted for CCITT No. 7. The MTP is responsible for transferring messages between signalling points in circuit-related applications. The objective of the MTP is to transfer messages without loss or duplication and deliver them to the intended destination in an error-free condition and in the sequence in which they were transmitted. The MTP provides a comprehensive signalling network that can meet demanding message-transfer requirements, even during fault conditions.

Level 1 of the MTP defines the physical, electrical and functional characteristics of the transmission link. Higher levels can thus be independent of the transmission system adopted to carry the signalling system.

Level 2 of the MTP defines the functions pertinent to a single signalling link. Information is transferred in signal units that can be message signal units (carrying user-part information), link-status signal units (reflecting the status of the signalling link) and fill-in signal units (used to maintain synchronisation). The most comprehensive signal unit is the message signal unit (MSU), comprising:

(*a*)   A flag delimiting the MSU
(*b*)   Sequence numbers used for error control
(*c*)   A length indicator
(*d*)   The service-information octet, indicating the appropriate user part or SCCP
(*e*)   The signalling-information field containing the user-part or SCCP information
(*f*)   Check bits used for error detection.

Two forms of error-correction mechanism are defined: the basic method and the preventive-cyclic-retransmission method.

Level 3 of the MTP defines the functions to provide a cohesive and comprehensive signalling network. Signalling-message-handling functions are used to analyse the originating-point code (OPC) and destination-point code (DPC), thus allowing messages to be routed to the correct destination-signalling point. The signalling-network-management functions define a range of features and facilities to control the flow of signalling messages through the network, e.g. a changeover feature allows messages to be re-routed away from a faulty signalling link.

The signalling-connection control part (SCCP) defines the functions, additional to the MTP, to meet the Layer 3/4 boundary requirements of the OSI 7-layer model. Thus, the combination of the MTP and SCCP provides a network service for higher layers. The SCCP is used in non-circuit-related applications.

The SCCP provides four classes of service to higher layers. Classes 0 and 1 are connectionless services in which data is transferred without the establishment of a signalling connection. In Classes 2 and 3, a signalling connection is established before data is transferred.

SCCP messages are carried in the signalling-information field (SIF) of the MTP. In this case, the SIF consists of:

(i)   A routeing label, containing the appropriate OPC and DPC,

(ii) A message-type field of 1 octet, uniquely defining the function of the message and

(iii) Various parameters. Parameters can be mandatory for a particular message type or optional. Mandatory parameters can be of fixed length or variable length. The formatting technique adopted for the SCCP is extremely flexible to enable future evolution to take place.

In connection-oriented services, a connection is first established using connection-request and connection-confirm messages. SCCP signalling information is then transferred between nodes using data messages. Messages are correlated by using local-reference numbers allocated by the nodes involved in the connection. This use of local-reference numbers, rather than the number of a traffic circuit, confirms the nature of SCCP as being non-circuit related. The connection is cleared by using released and release-complete messages. In connectionless services, SCCP signalling information is transferred in unitdata messages without establishing a connection.

The SCCP was defined internationally before widespread implementa-tions of interim systems took place. It is therefore possible to capitalise on the benefits of economies of scale in terms of development costs when implementing the transfer of non-circuit-related information. Because the SCCP, in conjunction with the MTP, offers a network service, it is possible that many higher-layer functions will be able to use the SCCP in the future, providing great flexibility for network operators.

5.9 References

1 CCITT Recommendations Q.701-Q.707: 'Message transfer part' (ITU,

Geneva)

2 LAW, B and WADSWORTH, C. A: 'CCITT Signalling System No. 7 in British Telecom's Network: the message transfer part', *British Telecom Engineering Journal*, 1988, **7**, P1, pp. 7–12

3 CCITT Recommendation Q.706: 'Message transfer part signalling performance' (ITU, Geneva)

4 CCITT Recommendation Q.703: 'Signalling link' (ITU, Geneva)

5 CCITT Recommendation Q.704: 'Signalling network functions and messages' (ITU, Geneva)

6 CCITT Recommendation Q.711: 'Functional description of the signalling connection control part (SCCP) of Signalling System No. 7' (ITU, Geneva)

7 CCITT Recommendation Q.713: 'Signalling connection control part (SCCP) formats and codes' (ITU, Geneva)

8 CCITT Recommendation Q.714: 'Signalling connection control part procedures' (ITU, Geneva)

Chapter 6

CCITT No. 7 user parts

6.1 Introduction

The user parts of CCITT No. 7 utilise the message-transfer part (MTP) to transfer signalling messages through the signalling network. Whereas the MTP provides a comprehensive transfer mechanism, including a dynamic routeing capability, the MTP cannot interpret the meaning of the Level-4 messages being transferred. It is the user part that defines the meaning of the messages that are being transferred and determines the sequence in which messages are sent. It is the user part that also interacts with the call-control function within an exchange to establish an overall controlling mechanism for calls. Three user parts have been specified for CCITT No. 7: the telephone user part (TUP) [1], the ISDN user part (ISUP) [2] and the data user part (DUP) [3]. All three are defined primarily to control the establishment and release of traffic circuits and they are, therefore, circuit-related in design.

In telephony, calls can be routed over analogue or digital transmission links and can be switched by analogue or digital exchanges. For integrated services digital network (ISDN) calls, there is a basic requirement to use digital transmission links and digital exchanges, thus ensuring that there is a digital connection (e.g. 64 kbit/s) from one customer to another.

The TUP was designed primarily to control the set-up and release of telephone calls. In addition to the control of basic telephony, the TUP defines the procedures and formats for extra features termed 'supplementary services'. An example of a supplementary service is when a called customer arranges for incoming calls to be diverted to another customer. To implement such supplementary services, the user parts must define the procedures that are adopted by participating exchanges. The TUP also provides a very basic ISDN capability by allowing a calling customer to request that a call is routed over digital transmission links and through digital exchanges. The TUP is described in Sections 6.2 to 6.5. The modern trend is to provide ISDNs and the ISUP is designed primarily to control the set-up and release of calls in such networks. The procedures and formats of the ISUP are optimised accordingly. Because of the flexible nature of ISDNs, the supplementary services defined within the ISUP are more comprehensive, and use more advanced techniques, than those defined for the TUP. More-comprehensive methods of selecting appropriate transmission links and exchanges are incorporated in the ISUP. The ISUP is described in Sections 6.6 to 6.8.

The evolution of the TUP to handle basic ISDN features, at the same time as the ISUP was being specified, represents a breakdown in effective international standardisation. Some operating companies are implementing early ISDNs using an enhanced version of the TUP, whereas others are adopting the ISUP. Hence, whilst the ISUP will be used universally in the long term in ISDNs, implementation in the short term will vary according to the policy of individual operating companies. This situation results from:

(a)   Lack of international agreement at the early stages of ISDN development on an evolutionary path

(b)   The slow nature of agreement of international standards.

If the full advantages of standardisation are to be gained in the future, it is essential that a single evolutionary path is defined and that appropriate standards are developed quickly enough to prevent interim implementations being needed.

The data user part (DUP) was defined in the early development of CCITT No. 7 to control the establishment and release of circuit-switched data calls. The implementation of the DUP is not extensive, with only a few network operators having implemented dedicated circuit-switched-data networks. Requirements for data networks will in future be handled by the ISUP, with the result that the DUP is unlikely to be used extensively in telecommunications networks. The DUP is therefore not described in this book: refer to Reference 3 for details of the specification.

6.2 Telephone user part

The telephone user part (TUP) defines the messages and procedures for telephone call-control signalling. It is designed for use in international telephone networks and can be used in national networks. National options are specified to allow for special circumstances in national networks. The TUP is designed to cover all telephone applications, including use over satellites.

The TUP is described in terms of its formats [4] and procedures [5] in Sections 6.3 to 6.5.

6.3 TUP formats

6.3.1 General
TUP messages are carried by the MTP in message-signal units, the general

format of which is described in Section 5.4.1 and Fig. 5.2. The format of a TUP message-signal unit is shown in Fig. 6.1. The functions of the flag, error-correction field, length indicator, service-information octet and check bits are described in Section 5.4. The user-part information is included in the signalling-information field (SIF) of the message. The SIF contains a label, a heading code and one or more information elements (which can be mandatory or optional, according to the type of message). The length of the signal unit can be fixed or variable.

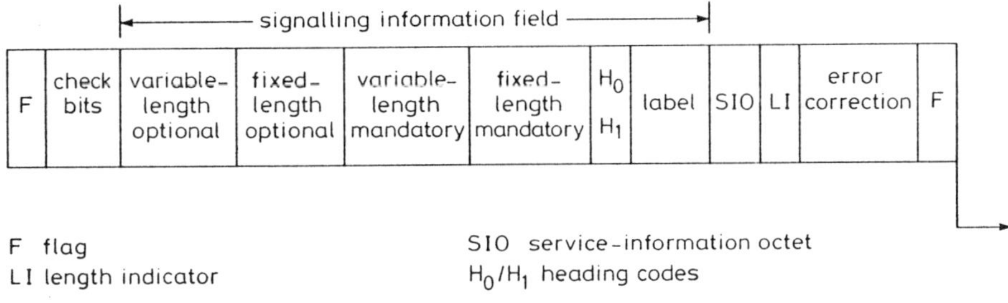

Fig. 6.1   TUP message – signal unit

The label is the information that the TUP uses to identify the speech circuit to which a message refers. The label consists of the routeing label (illustrated in Fig. 5.4) and the circuit-identification code (CIC). The routeing label specifies the identity of the originating and destination exchanges. The CIC is a number that identifies the speech circuit to which the message refers. In the TUP, the CIC includes the 4-bit signalling-link selection (SLS) field described in Section 5.5.2.

All TUP messages contain a heading code, consisting of an H0 field and an H1 field. The H0 field identifies to which general category (message group) a message belongs. For example, a message containing address information (digits dialled by the calling customer) belongs to the category 'forward-address message'.

The meaning of the H1 field depends on the complexity of the message. For a simple message, the H1 field contains enough information to define fully the meaning of the message. For example, if a telephone conversation has been completed and the calling customer clears (i.e. releases) the call, the H1 field consists of information instructing exchanges to clear the speech path that is being used. For a complex message, the H1 field contains information that specifies the format of the remainder of the message. For example, in a message setting-up a call, the H1 field indicates the type of information that is included in the rest of the SIF.

After the label and heading codes, the SIF generated by the TUP is divided into a number of subfields. Mandatory subfields are compulsory for a given message type and they appear in all messages of that type. Optional subfields can be present, but are not compulsory, for a given message type. Mandatory and optional subfields can be of fixed length or variable length. Fixed-length subfields have the same number of bits in each message of a given type. Variable-length subfields can vary in size for a message of a given type. The length of a variable-length subfield is specified by a 'length indicator'. The order of subfield transmission in the SIF is:

(i)   Fixed-length-mandatory subfields
(ii)   Variable-length-mandatory subfields
(iii)   Fixed-length-optional subfields
(iv)   Variable-length-optional subfields

A wide range of message formats is defined by CCITT [4]. To illustrate the principles involved, examples of formats of commonly-used messages are described in Sections 6.3.2 to 6.3.5.

6.3.2 Initial-address message (IAM)

The IAM is the first message to be sent during the set-up of a call. It contains the required address (e.g. digits dialled by the calling customer) and other information needed for routeing purposes. The format of the IAM is shown in Fig. 6.2.

| address signals | number of address signals | message indicators | calling-party category | 0001 | 0001 | label |
| | | | | H_1 | H_0 | |
| n x 8 | 4 | 12 | 8 | 4 | 4 | 40 | bits

Fig. 6.2   Format of TUP initial-address message
The calling-party category includes 2 spare bits
Source: CCITT Recommendation Q.723

The IAM belongs to a message group termed 'forward-address messages' and the HO field is coded accordingly (0001). The IAM is a complex message: the H1 code (0001) identifies the message as an initial-address message and determines the format of the rest of the SIF. The calling-party category indicates the type of customer initiating the call, e.g. ordinary customer or operator. The message indicators give a variety of information pertinent to the call, e.g. whether or not a satellite is being used to provide one of the transmission links. The address digits dialled by the calling customer are included in the IAM after the message indicators. A subfield precedes the digits to indicate the number of address digits included in the message.

6.3.3 Address-complete message (ACM)

The ACM is sent by the destination exchange to the originating exchange to indicate that all the address digits required to identify the called customer have been received. The format is shown in Fig. 6.3.

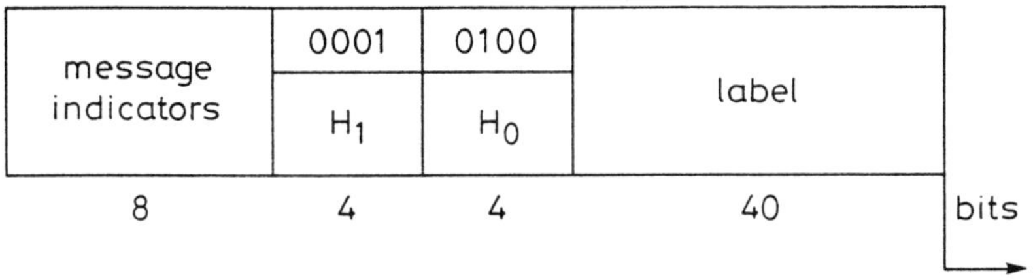

message indicators	0001	0100	label
	H_1	H_0	
8	4	4	40 bits

Fig. 6.3   Format of TUP address – complete message
Source: CCITT Recommendation Q.723

The ACM belongs to the group of messages termed 'successful backward set-up information messages'. The H0 is coded 0100 in accordance with the message group. The H1 code of 0001, in conjunction with the HO code, identifies the message as an ACM. The message indicators give general information about the characteristics of the established call, e.g. whether or not echo suppressors are involved with the speech path.

6.3.4 Answer signal (ANS)

The answer signal is sent by the destination exchange to indicate that the call has been answered by the called customer. The answer signal belongs to the group of messages termed 'call supervision', with the general format shown in Fig. 6.4. The ANS is a simple message: the message consists only of the H0/H1 combination, without further indicators. The H0 code for the call supervision message is 0110 and the H1 code for a normal answer message is 0000.

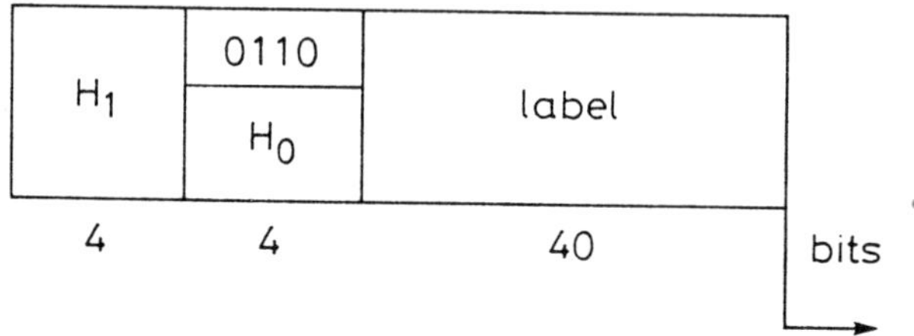

Fig. 6.4   Format of TUP call-supervision message

6.3.5 Clear-forward (CLF) signal

The clear-forward signal is sent by the originating exchange when the calling customer indicates that the call has been completed. The CLF is a call-supervision signal with the same general format as the answer signal shown in Fig. 6.4. Again, the CLF is a simple message and the H0/H1 combination is sufficient to define the message. For CLF, the H1 code is 0100.

6.4 TUP procedures

6.4.1 Basic call set-up

A basic call set-up sequence under normal operating conditions is shown in Fig. 6.5. The originating exchange initiates the call by sending an initial-address message (IAM) to the intermediate exchange. The direction in which the IAM is transmitted (originating exchange to intermediate exchange) determines the 'forward direction' of the call. The opposite direction is termed the 'backward direction'. The IAM contains all the information relating to the characteristics of the required speech path. It may also contain all the address digits required to identify the called customer: this is termed 'en bloc' operation. Alternatively, the IAM may contain only sufficient address digits to route the call to the intermediate exchange: this is termed 'overlap' operation. In both cases, sending an IAM results in the seizure of an appropriate speech path between the originating and intermediate exchanges. Seizure indicates that the speech path is reserved. At this stage, the originating exchange does not through-connect the speech path to the calling customer. The intermediate exchange analyses the received IAM and selects an appropriate speech path to the destination exchange. The IAM is forwarded to the destination exchange over an appropriate signalling link.

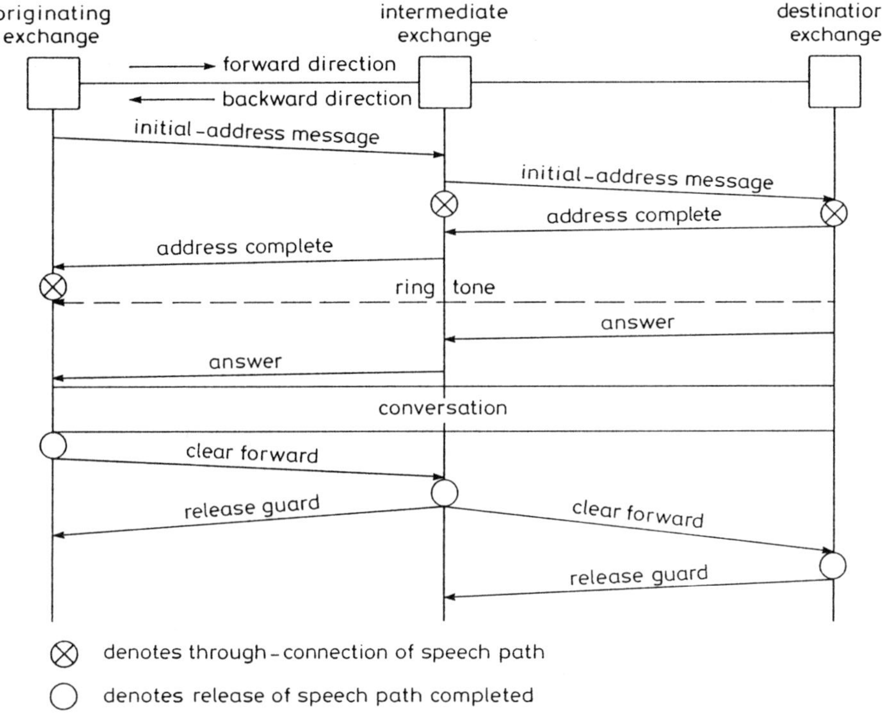

Fig. 6.5   Basic call set-up and release procedures for the TUP

In en bloc operation, all the relevant information is included in the IAM to reach the called customer. In this case, the destination exchange returns an address-complete message (ACM) to confirm receipt of the called customer's number. In overlap operation, only some of the address digits are included in the IAM. In this case, the originating exchange supplies the extra digits in one or more subsequent-address messages (not shown in Fig. 6.5). When all address digits are received at the destination exchange, the ACM is returned to the originating exchange. In both the en bloc and overlap methods of operation, the receipt of the ACM at the originating exchange causes it to connect through the speech path to the calling customer.

Whilst returning the ACM to the originating exchange, the destination exchange commences ringing the called customer. The destination exchange also returns ring tone over the allocated speech path to the calling customer. When the called customer answers the telephone, ringing and ring tone are stopped and an answer message is returned to the originating exchange. It is usual for the originating exchange to commence charging upon receipt of the answer message. At this stage, conversation may take place.

6.4.2 Call release

The basic call-release procedures are illustrated in Fig. 6.5. Upon completion of the call, either the calling customer or the called customer can initiate the release of the speech path. However, only the calling customer can initiate an immediate release of the speech path. If the calling customer initiates the release procedure, the originating exchange commences to release the speech path. When the speech path is released, the originating exchange sends a clear-forward (CLF) message to the intermediate exchange. Upon receipt of the CLF message from the originating exchange, the intermediate exchange commences the release of the speech path. When the release is complete, the intermediate exchange sends a CLF message to the destination exchange and a release guard (RLG) message to the originating exchange. Upon receipt of the RLG message at the originating exchange, the speech path is made available for new traffic. Upon receipt of the CLF message at the destination exchange, the destination exchange releases the speech path. When this is complete, it returns a RLG message to the intermediate exchange. Upon receipt of the RLG message at the intermediate exchange, the speech path is made available for new traffic.

In telephony networks, the normal method of operation is termed 'calling-party release'. This means that the calling customer is deemed to control, and have the authority to release, the call. This approach differs from circuit-switched data networks, in which a 'first-party release' method of operation is generally used: in this case, both the calling and called customers have the authority to release a call. The TUP operates the calling-party release method of operation, using the clear-forward sequence described above. However, there is a need to safeguard the calling customer from unintentionally prolonging a call (e.g. due to faulty equipment preventing the recognition of a call-release request). For this purpose, a clear-back sequence is also specified. In this case, if the called customer ends the conversation, the destination exchange generates a clear-back (CLB) message and sends it to the intermediate exchange. The intermediate exchange examines the CLB message and, without further action, passes it back to the originating exchange. After receipt of the CLB message, the originating exchange commences a timing period of 1–2 minutes. If the calling customer clears down during this period, then the clear-forward sequence described above is enacted immediately. If no indication is received from the calling customer, the originating exchange waits until expiration of the timing period before commencing the clear-forward sequence.

6.4.3 Abnormal conditions

The specification of a complex signalling system needs to cover abnormal conditions. The aim is to ensure that, even in abnormal circumstances, appropriate action is taken by the telephone network to overcome the problem and maintain service. The TUP covers many such conditions [5]; examples are 'reset', 'dual seizure' and 'abnormal release'.

The reset procedure is used when the status of a speech circuit, or a number of circuits, becomes unclear to an exchange (e.g. due to memory mutilation within the exchange). In this case, procedures are defined for the exchanges at each end of a circuit to reset the circuit to the idle condition. If a small number of circuits are affected, reset messages are used for each circuit. If a large number of circuits are affected, a group-reset message may be applied to reset all the circuits with one message.

Dual seizure is when two inter-connected exchanges choose the same speech circuit for two different calls at approximately the same time. The risk of dual seizure is minimised by applying different algorithms for choosing speech paths at each end of the transmission link. In this way, the likelihood of two exchanges choosing the same speech circuit is reduced. To resolve the contention when a dual seizure does occur, each circuit is nominally controlled by one exchange. The IAM sent by the controlling exchange for a particular circuit is processed normally by the non-controlling exchange. The IAM sent by the non-controlling exchange for the circuit is disregarded by the controlling exchange. Upon detection of dual seizure at the non-controlling exchange, an automatic repeat attempt is initiated for another speech path.

A series of rules is defined covering the actions to be taken at each exchange upon receipt of an unexpected message. For example, if an exchange receives a clear-forward message relating to an idle circuit, the exchange will acknowledge the signal with a release guard message. Receipt of many other unexpected messages causes the return of a reset message.

6.5 TUP supplementary services

6.5.1 General

The TUP defines procedures and formats for supplementary services to provide features additional to basic call set-up and release. The procedures for supplementary services are very complex owing to:

(*a*)   The need to specify the procedures in sufficient detail without imposing undue constraints upon the network operator,

(*b*)   The complex nature of some supplementary services

(*c*)   The need to cover abnormal conditions and failures at each stage of each supplementary service

(*d*)   The changing nature of many supplementary services
(*e*)   The impact of combining supplementary services.

Detailed procedures are given in the specification [5] and will not be discussed here. However, below is a brief description of some of the more-common supplementary services specified for the TUP.

6.5.2 Closed-user group

The closed-user group (CUG) facility allows customers to form groups or clubs. Members of the group can call each other, subject to restrictions depending upon the type of CUG. Non-members cannot call (or be called by) members, unless specifically allowed by the CUG type.

The CUG facility is implemented using an interlock code. During call set-up, a validation check is performed to ensure that the calling and called customers belong to the same CUG, as verified by the interlock code. The data to verify interlock codes can be held at local exchanges (decentralised version of CUG) or at dedicated points in the network (centralised version of CUG). Only the decentralised version of CUG is specified for the TUP by CCITT [5].

6.5.3 User access to calling-line identification

This facility allows the presentation of the calling customer's telephone number to the called customer, before the called customer answers the telephone. The calling customer can prevent the presentation of the calling-customer number by invoking a further facility termed 'calling-line-identity presentation restriction'.

There are two ways of transferring the calling customer's number through the network. The number can either be included in the initial-address message generated by the originating exchange or the number can be requested by the destination exchange when an IAM is received that does not include the appropriate information.

6.5.4 User access to called-line identification

This facility is used by the calling customer to confirm to which telephone number a call is connected. An indicator is included in the IAM requesting the called-customer address. The address is returned by the destination exchange in response to the indicator.

6.5.5 Redirection of calls

The redirection facility allows calls to a particular telephone number to be redirected to another predetermined number. Procedures are included to allow a called customer to reject a redirected call. To avoid numerous redirections, resulting in a continuous loop, only one redirection is allowed per call.

When the call is delivered to the original-destination exchange (i.e. the exchange to which the original-called customer is connected), the original-destination exchange recognises that a redirection is required and initiates an IAM to the new-destination exchange (i.e. the exchange to which the call is being redirected). The IAM contains an indicator that the call has been redirected. An ACM is returned from the new-destination exchange, including a parameter to indicate that the call has been redirected. Subsequent set-up procedures are as for a basic call.

6.5.6 Digital connectivity

A calling customer can request that the speech path to be established through the network should be digital. An indicator is included in the initial-address message to allow analysis at each exchange. If it is not possible to provide digital equipment for each connection, the call is rejected and the originating exchange is informed that the call cannot be completed.

6.6 ISDN user part

The ISDN user part (ISUP) defines the messages and procedures for the control of switched services in ISDNs. The ISUP covers both voice (e.g. telephony) and non-voice (e.g. circuit-switched data) applications. As telecommunications networks evolve towards ISDN, the ISUP will obviate the need for the TUP and DUP.

The TUP was developed primarily for the international network, with allowances for national networks being made as necessary. The ISUP was developed from the outset for operation in both national and international networks. Hence, the long-term adoption of ISUP will achieve the aim of reducing signalling interworking costs between the international network and national networks.

The ISUP contains all the functions of the TUP, but the functions are achieved in a more-flexible manner. Extra features are also provided; one major feature is end-to-end signalling, which allows two exchanges to communicate without intermediate exchanges analysing the messages. Further detail is given in Section 6.8.4.

The principles of the ISUP formatting [6] are described in Section 6.7 and the procedures for providing calls [7] are described in Section 6.8.

6.7 ISUP formats

6.7.1 Format principles

The principles adopted for TUP formatting were considered to be restrictive when applied in an ISDN environment. Each time a new TUP message is required or an existing message is modified, the specification needs to be revised. By placing more emphasis on the use of variable and optional fields, within a framework defined by the specification, the ISUP can be more responsive to changing requirements. In addition, modern processing techniques can be applied more efficiently using a repetitive approach in which the order of fields is: parameter name, length of parameter and specific information. In this respect, the formatting principles adopted for the ISUP are similar to those adopted for the SCCP (Section 5.6.3). Whereas the SCCP is non-circuit-related in nature, and therefore uses a local reference to identify a particular transaction, the ISUP maintains the circuit-related approach for transaction identification, i.e. the number of the circuit is used in the message to identify information pertaining to that circuit. Hence, a circuit-identification code (CIC) is used in the ISUP, in common with the TUP (Section 6.3.1).

The ISUP messages are carried in the SIF of message-signal units, as illustrated in Fig. 5.2. The signalling information field consists of:

(*a*)   The routeing label
(*b*)   A circuit-identification code
(*c*)   The message type
(*d*)   Parameters

The parameters are divided into the mandatory-fixed part, the mandatory-variable part and the optional part, as illustrated in Fig. 6.6. The routeing label is described in Section 5.5.2. The circuit-identification code (CIC) indicates the number of the speech circuit between two exchanges to which the message refers.

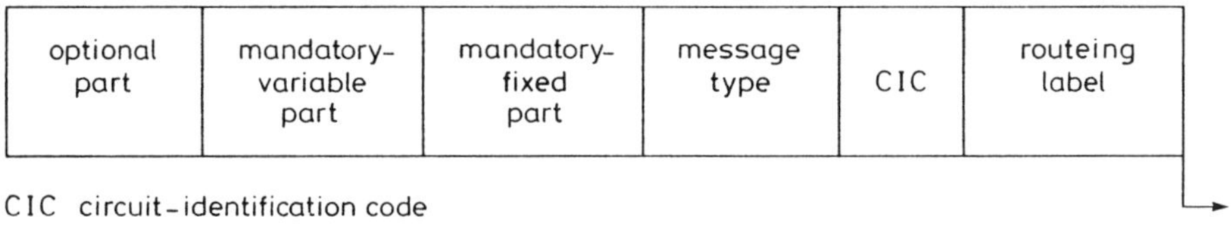

optional part	mandatory-variable part	mandatory-fixed part	message type	CIC	routeing label

CIC   circuit–identification code

Fig. 6.6   Signalling – information field for the ISUP

The message-type code consists of a one octet field and is mandatory for all messages. The message-type code uniquely defines the function and the general framework of each ISUP message.

Each message includes a number of parameters. Each parameter has a name which is coded as a single octet. The length of a parameter can be fixed or variable. Parameters are categorised as:

(i)   Fixed mandatory
(ii)   Variable mandatory
(iii) Optional.

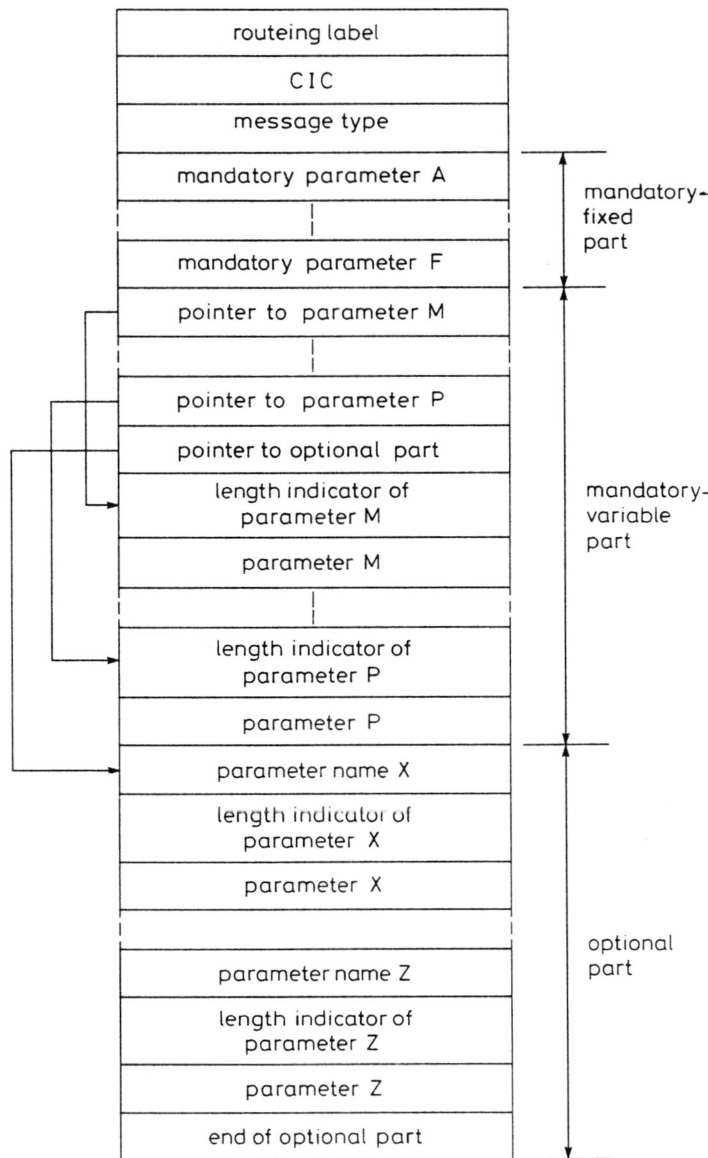

Fig. 6.7 Structure of parameters in the ISUP
Source: CCITT Recommendation Q. 763

Fixed-mandatory parameters are mandatory for a given message type and are of fixed length. The position, length and order of the parameters are uniquely defined by the message type, so parameter names and length indicators are not included in the message.

Variable-mandatory parameters are mandatory for a given message type and of variable length. A 'pointer' is used to indicate the beginning of each parameter. A pointer is an octet that can be used during processing of the SIF to find a particular piece of information: this avoids having to analyse the whole message to find one piece of information. The name of each parameter is implicit in the message type, so parameter names are not included in the message. A pointer is included to indicate the beginning of the optional part of the SIF.

Optional parameters may or may not occur in any particular message type. Each optional parameter includes the name (one octet) and a length indicator (one octet) before the parameter contents.

An illustration of the message format of the ISUP, giving the structure of the parameters, is shown in Fig. 6.7.

6.7.2 Examples of message formats

The CCITT specification of the ISUP gives a range of message types and parameters [6]. Examples of some message types to illustrate the principles are:

(*a*)   Initial-address message (IAM)
(*b*)   Address-complete message (ACM)
(*c*)   Answer (ANS) message
(*d*)   Release (REL) message

The initial-address message (IAM) is the first message to be sent during call set-up. It contains the address digits (e.g. digits dialled by the customer to route the call) and it results in a seizure of a circuit by each exchange. The general format of the IAM is shown in Table 6.1, with examples of optional parameters. The message type of the IAM is coded 00000001.

The address-complete message (ACM) is sent by the destination exchange to indicate successful receipt of sufficient digits to route the call to the called customer. The general format of the ACM is shown in Table 6.2, with examples of optional parameters. The message type of the ACM is coded 00000110.

The answer (ANS) message is sent by the destination exchange to indicate that the called customer has answered the call. The format of the answer message is shown in Table 6.3. The message type of the ANS message is coded 00001001.

Table 6.1 Format of initial-address message

Parameter	Type	Length (octets)	Description
Nature of connection	F	1	Gives status of connection being established, e.g. satellite included/not included, echo suppressor included/not included
Forward-call indicators	F	2	Indicates capability of the connection, e.g. end-to-end possible/not possible, ISUP available/not available throughout connection
Calling-party category	F	1	States whether calling customer is subscriber or operator (plus language)
Transmission-medium requirements	F	1	Requests type of transmission, e.g. 64 kbit/s connection
Called-party number	V	4–11	Number of called customer (e.g. dialled digits)
Calling-party number	O	4–12	Number of calling customer
User-to-user information	O	3–131	Capacity can be provided to allow customers to send data to each other during the set-up sequence

F : Fixed mandatory, V : Variable mandatory, O : Optional

The release (REL) message can be sent by the originating or destination exchanges to clear-down the traffic circuit. Either the calling or the called customer can initiate the release of the circuit. The general format of the REL message, with examples of optional parameters, is shown in Table 6.4. The release message has a message type coded 00001100.

6.7.3 Format overhead

The ISUP makes extensive use of the optional parameter fields, thus increasing the flexibility available to network operators. However, such flexibility can increase the processing overhead within any exchange that analyses the information. The IAM specified by CCITT can contain up to 14 optional parameters, including up to 131 octets of user-to-user information. Hence, two factors could cause a processing overhead:

(*a*)   The sheer size of some ISUP messages could become a problem if too many optional fields are included within one message and

Table 6.2 Format of address-complete message

Parameter	Type	Length (octets)	Description
Backward-call indicators	F	2	Indicates eventual capability of the connection, e.g. end-to end signalling available/ not available.
Optional backward-call indicators	O	3	Additional indicators for use with, for example, supplementary services
User-to-user information	O	3–131	As described for IAM

Table 6.3 Format of answer message

Parameter	Type	Length (octets)	Description
Backward-call indicators	O	4	As described for ACM
User-to-user information	O	3–131	As described for IAM

Table 6.4 Format of release message

Parameter	Type	Length (octets)	Description
Cause indicators	V	≥ 3	Gives reason for the release of the call, eg. calling/called customer cleared-down
User-to-user information	O	3–131	As described for IAM

(*b*)   The flexible approach of optional fields in itself requires extra
processing to determine what information has been provided in a
particular message. Provided that these features are treated carefully,
the ISUP formatting technique is extremely flexible and will permit
evolution to handle future requirements.

6.8 ISUP procedures

6.8.1 Basic call set-up and release

The procedures for basic call set-up and release are illustrated in Fig. 6.8.
Upon receipt of a request to set up a call from the calling customer, the
originating exchange analyses the routeing information and formulates an
initial-address message (IAM). Analysis of the called-party number allows
the originating exchange to determine to which exchange to route the call.
In this example, it is recognised that the call must be routed to the
intermediate exchange. Examination of the information supplied by the
calling customer indicates the type of connection required (e.g. 64 kbit/s
connection through the network). This information is used to select an
appropriate circuit and to provide relevant data in the IAM (e.g. set
indicators in the transmission-medium-requirement parameter to allow the
intermediate exchange to select an appropriate circuit). The IAM is sent to
the intermediate exchange and the corresponding speech path is switched
through in the backward direction to the calling customer. Switching
through in the backward direction only at this stage allows the calling party
to hear tones provided by the network, but prevents the calling party from
sending information on the speech path. If en bloc operation is used, all the
address digits needed for routeing to the called customer are included in the
IAM. If overlap operation is used, the IAM is sent when sufficient digits have
been received to route to the intermediate exchange, with other digits being
sent across the network in subsequent-address messages.

The intermediate exchange receives the IAM and analyses the information contained. By examination of the address digits, the intermediate
exchange determines the routeing to the destination exchange. By examination of the other information in the IAM (e.g. the transmission-medium-
requirement parameter), an appropriate speech path is selected (e.g. a
64 kbit/s circuit). The IAM is sent to the destination exchange and the
speech path is through-connected.

When the IAM is received at the destination exchange, it is analysed to
determine the called customer. The information in the IAM is examined to
determine if further information is required from the originating exchange
before connection to the called customer takes place (e.g. the calling-
customer address might be required, but it might not be included in the

IAM). If further information is required, an end-to-end message is returned to the originating exchange stating the requirement. Note that the intermediate exchange does not need to analyse the end-to-end message and it is transported transparently. The originating exchange supplies the relevant information by returning an end-to-end message.

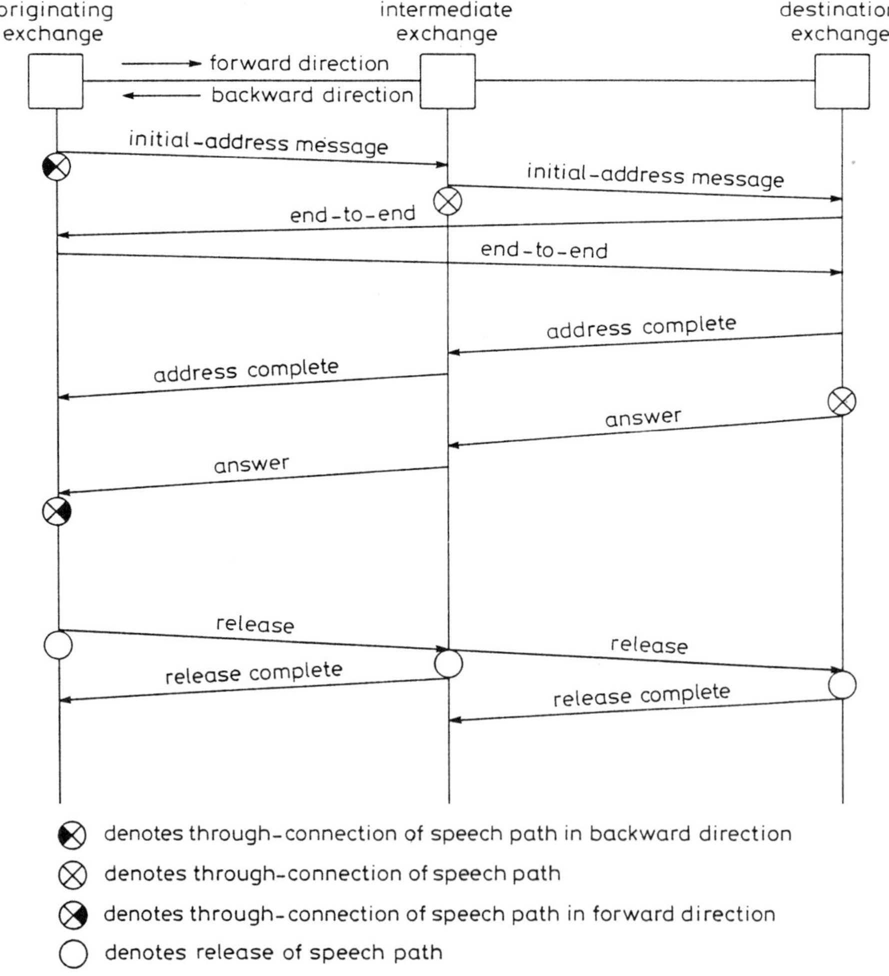

Fig. 6.8 ISUP basic call set-up and release

Upon receipt of all relevant information at the destination exchange, the called customer is informed that a call is being established and an address-complete message is sent from the destination exchange to the intermediate exchange. The address-complete message is subsequently sent to the originating exchange. Receipt of the address-complete message at any exchange indicates successful routeing to the called customer and allows specialised routeing information to be deleted from the memory associated with the call in exchanges involved in the connection.

When the called customer answers the call, the destination exchange switches through the speech path and returns an answer message to the intermediate exchange. The intermediate exchange returns the answer message to the originating exchange. Upon receipt of the answer message, the originating exchange switches through the speech path in the forward direction. The calling and called customers are now connected, charging can commence and conversation or data transfer can occur.

In some circumstances, influenced by the type of terminal used by the called customer, a response can be received from the called customer before the address-complete message is sent to the originating exchange. In this case, a 'connect' message is sent to the originating exchange. The connect message combines the functions of the address-complete and answer messages and obviates the need for these two messages.

Unlike the TUP, either the calling or called customer can initiate immediate release of the connection; i.e. the ISUP adopts the first-party release method of operation, thus catering for data applications. In Fig. 6.8, the calling customer requests disconnection from the originating exchange. The originating exchange commences the clear-down of the connection and sends a release message to the intermediate exchange. The intermediate exchange sends a release message to the destination exchange and commences clear down of the speech path. When the speech path has cleared-down and is ready for new traffic, the intermediate exchange returns a release-complete message to the originating exchange. Similarly, upon receipt of the release message, the destination exchange commences clear-down of the speech path. When the clear-down is complete, the destination exchange returns a release-complete message to the intermediate exchange.

One concept influencing the design of the release procedure is the need to ensure that either customer can release the connection as soon as possible. Whereas the TUP transfers the clear-forward message upon completion of clear-down, the ISUP transfers the release message immediately, thus increasing the speed of transmission through the network. The original specification of the ISUP defined a three-message-release sequence involving the release, released and release-complete messages. This was replaced by the procedure described above to achieve more commonality with the SCCP release procedures.

6.8.2 Additional features

The ISUP is designed to operate in ISDNs, so features additional to those provided by the TUP are required to ensure that both telephony and data applications can be handled. Examples of such features are:

(*a*) Suspend/resume
(*b*) In-call modification

The suspend/resume feature allows a calling or called customer to suspend communication for a period of time during the conversation/data phase of a call. The feature can be used to allow a customer to change the terminal equipment that is being used, or to change locations within the customer's premises, without releasing the call. The feature is invoked by either customer sending a 'suspend-request' message, which is passed through the network to the other party. When communication is required again, either customer can send a 'resume' message to the other party. One exchange in the network commences a timer upon receipt of a suspend-request message to prevent unintentionally-long suspensions.

The in-call modification feature allows the calling or called customer to modify the characteristics of the connection during the conversation/data phase of the call. An example of the application of this feature is when the calling and called customers wish to change from using data over the connection (64 kbit/s data) to using speech. During the connection establishment, the IAM would have included parameters to indicate that a data call was required; hence echo suppressors would not have been connected. If there is a wish to change to speech during the call, echo suppressors may be required. The in-call-modification procedure allows the echo suppressors to be provided during the call by using a 'call-modification-request' message. When each exchange has made the necessary modifications, the last exchange in the chain returns a 'call-modification-complete' message, thus confirming that speech can commence. Complex procedures exist to ensure that failure to provide appropriate changes to an in-call-modification request are recognised and reported to the calling and called customers.

6.8.3 Abnormal conditions

The criteria and actions for many abnormal conditions are similar to the TUP procedures described in Section 6.4.3, including reset, dual seizure and abnormal release. However, the procedures in the ISUP are more comprehensive than those for the TUP to cater for the more-flexible formats. There is also recognition in the ISUP specification that not all exchanges in a network operate with the same version of CCITT No. 7. It is not possible to upgrade all exchanges simultaneously: at any time, therefore, some exchanges operate with one version of CCITT No. 7, whereas others operate with a more modern (enhanced) version.

Procedures are defined in the ISUP to cater for these circumstances. Consider two exchanges provided with ISUPs, as shown in Fig. 6.9. Exchange A has been upgraded to operate an enhanced version of the ISUP providing new features. Exchange B has not yet been upgraded and is operating the standard version. If Exchange B receives a message that it does not understand, then it returns a 'confusion' message, with a parameter called 'unrecognised message', to Exchange A. This indicates that Exchange B cannot process the call. Exchange A can therefore:

(*a*)   Send an alternative message that Exchange B can understand, if this is possible

(*b*)   Re-route the call to another exchange, seeking an enhanced version of ISUP

(*c*)   Inform the calling customer that the new feature is not yet available on the route requested.

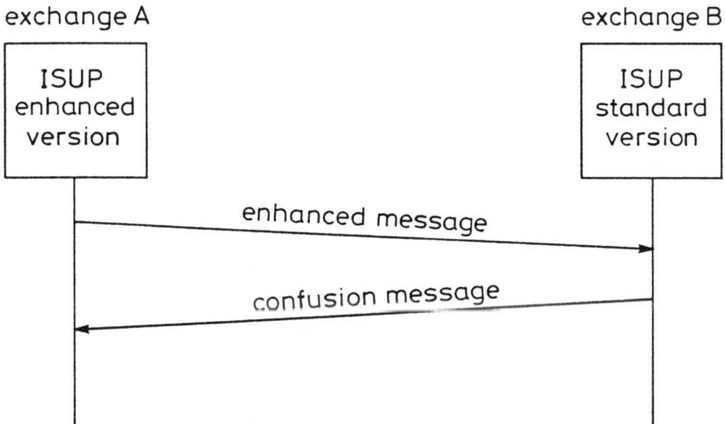

Fig. 6.9   ISUP interworking of versions

Similar procedures are defined if unrecognised parameters or parameter values are received at Exchange B. If it is possible to continue the call without the unrecognised information, Exchange B processes the call. If it is not possible to continue the call without the unrecognised information, procedures are defined to release the call.

These procedures for handling unrecognised messages and information will become increasingly important for the efficient operation of networks. The need for rapid changes to meet customer demands means that more versions of the ISUP will be implemented, resulting in a greater need for an efficient method of allowing versions to interwork.

6.8.4 End-to-end signalling

End-to-end signalling allows exchanges to send and receive signalling information without the signalling information being analysed by intermediate (e.g. trunk) exchanges. End-to-end signalling is typically used between local exchanges to transfer specialised information about a feature requested by the calling or called customer. In this case, the end-to-end signalling is routed via trunk exchanges, but the trunk exchanges do not analyse the contents of the messages. In this context, the local exchanges are termed 'end-points'. Two forms of end-to-end signalling are specified:

(*a*)   Pass-along
(*b*)   SCCP method.

The pass-along method of end-to-end signalling makes use of the routeing information established for a call. When the ISUP establishes a telephone or data call, the number dialled by the calling customer is translated into routeing information for use in the signalling network. The routeing information is the routeing label plus the CIC (see Section 6.7.1). Routeing information is held by each exchange in the connection for the duration of the call. A message type termed 'pass-along' is defined. When, during a call, a trunk exchange receives a pass-along message, it uses the routeing information already available to transfer the message to the next exchange, avoiding the analysis of the end-to-end information contained within the pass-along message. Only the local exchanges (end-points) sending and receiving the end-to-end information need to analyse the full message.

The SCCP method of end-to-end signalling uses the SCCP (Section 5.6) to transfer signalling information. Two forms of information transfer are specified: connectionless service and connection-oriented service.

In the connectionless service, an ISUP message (typically an IAM) being transferred from an originating exchange to a destination exchange includes a 'call reference'. The inclusion of a call reference is an implicit indication to the destination exchange (receiving end-point) that a connectionless SCCP exchange of information is required. Upon receipt of a call reference at the receiving end-point, a corresponding call reference is returned to the originating exchange (sending end-point) in an address-complete message. This exchange of call references permits the transfer of unitdata messages using the SCCP. The call references are used to correlate the unitdata messages with the call.

In the connection-oriented service, a SCCP connection-request (CR) message is embedded within an ISUP message. If end-to-end signalling is required at the same time as establishing an ISUP call, the CR message is embedded within an IAM (other types of ISUP message can be used if a call already exists). Receipt of an IAM with an embedded CR message at a destination exchange, indicates that the originating exchange wishes to establish an end-to-end connection. The CR message is passed by the ISUP at the destination exchange to the SCCP. The SCCP at the destination exchange then responds directly to the SCCP at the originating exchange with a connection-confirm (CC) message. Information transfer then takes place with data messages using normal SCCP procedures.

The end-to-end techniques described above represent the ability to establish a logical connection between end-points (i.e. an efficient means of exchanging signalling information between end-points), as well as the physical connection of circuits. This is a very flexible and powerful tool and it is expected that these techniques will be used a great deal in the future, particularly to provide advanced features for customers.

6.8.5 Supplementary services

The ISUP defines the formats and procedures for the following supplementary services [8]:

(*a*)   User-to-user signalling,
(*b*)   Closed-user group,
(*c*)   Calling-line identification,
(*d*)   Direct-dialling in and
(*e*)   Call forwarding.

Some of these features (e.g. calling-line identification) are similar to those adopted for the TUP, whereas others (e.g. user-to-user signalling) are only defined for the ISUP.

User-to-user signalling allows the transfer of data between the calling and called customers through the signalling network. As described for end-to-end signalling, nodes in the network do not analyse the user-to-user data: it is transported transparently from one customer to another. For user-to-user signalling , even the local exchanges do not analyse the information.

There are three user-to-user signalling services. In Service 1, user-to-user information is included within the initial-address, address-complete, answer and release messages. In Service 2, user-to-user information is exchanged during call set-up between the address-complete (or call-progress) and answer messages. In Service 3, user-to-user information is exchanged during the conversation/data phase of a call using 'user-to-user information' messages. Each service can be applied independently to a call or they can be applied together.

The procedures for the closed-user-group facilty depend upon validation of authority for two customers to communicate, as explained in Section 6.5.2 for the TUP. This validation can be carried out in a decentralised method (in which the originating and destination local exchanges store authority information) or in a centralised method (in which a central database·stores the authority information). The TUP only specifies the former approach, but ISUP specifies both methods. This is because non-circuit-related procedures are required for the centralised approach and these are only specified for the ISUP.

The procedures for the decentralised method of CUG are very similar to those defined for the TUP. The procedures for the centralised method are based upon the originating and destination exchanges sending non-circuit-related messages to a central database, seeking confirmation that two customers are allowed to communicate within the terms of the CUG.

The procedures for calling-line identification (CLI) are very similar to those described for the TUP in Section 6.5.3. In the ISUP, there is more emphasis upon including the calling-line identity in the initial-address message.

Direct-dialling in (DDI) enables a call to be made to a customer connected to a private automatic branch exchange (PABX) without the intervention of a PABX operator. Procedures are defined in the ISUP to cover both analogue and digital DDI PABXs.

The basic procedures for call forwarding are similar to the redirection facility in the TUP. However, the ISUP call forwarding sequence can be initiated under differing circumstances, as follows:

(i)   For all calls when the facility is activated
(ii)   When the called customer is busy or
(iii)   When there is no reply from the called customer after a specified period.

In the ISUP, the parameters included in the IAM sent to the new destination exchange are extensive and it is possible to allow multiple call forwarding i.e. a call can be passed to successive customers if the call-forwarding facility is activated.

6.9 Chapter summary

User parts comprise Level 4 of the 4-level structure adopted for CCITT No. 7. The user parts use the MTP to transfer messages between signalling points, the messages being carried in the signalling-information field. The user parts define the meaning of the messages carried by the MTP.

The telephone user part (TUP) is defined primarily to set-up and clear-down telephone calls. As well as basic telephony, the TUP also defines procedures for some supplementary services e.g. closed-user group, redirection of calls and calling-line identification. The ISDN user part (ISUP) is designed for use within ISDNs and covers telephony and circuit-switched data. The ISUP also provides supplementary services and features additional to the TUP. The data user part was designed for use in circuit-switched-data networks, but such networks are rare and future data requirements will be handled by the ISUP.

The TUP signalling-information field consists of a label, a heading code and sub-fields. The label consists of the routeing label (including the OPC and DPC) and a circuit-identification code. The heading code defines the class of message and, for simple messages, defines the message-type. For complex messages, the heading code defines the format of the rest of the message. The sub-fields can be mandatory or optional. In both cases, sub-fields can be of fixed length or variable length.

Traffic circuits are established in the TUP by sending an initial-address message including the address of the called customer. An address-complete message indicates that sufficient address information has been supplied and an answer message indicates that the called customer has answered the call. The call can be cleared by the calling customer initiating a clear-forward message. The traffic circuit is not made available for another call until a release guard message is received in response to the clear-forward message.

The ISUP has a more flexible formatting technique than that adopted by the TUP, more reliance being placed upon variable-length and optional fields. The formatting technique for the ISUP is similar to that adopted by the SCCP. The ISUP signalling-information field consists of a routeing label, circuit-identification code, message type and parameters. The message type is one octet in length and uniquely defines the function of the message. The parameters can be:

(*a*)   Mandatory and fixed length
(*b*)   Mandatory and variable length
(*c*)   Optional.

The procedures for setting up a basic call are similar to those for the TUP, except that through connection of the traffic circuit occurs at different times. In addition, the ISUP permits the use of end-to-end signalling during call establishment. The end-to-end signalling can use a pass-along technique, making use of routeing data generated for the call, or it can use the SCCP. The release procedures for the ISUP use a release and a release complete message. The ISUP release procedures are faster than the TUP and they can be initiated by either the calling or the called customer.

The ISUP includes additional features for use in ISDNs, including a suspend/resume procedure, which allows the calling and called customers to suspend the call temporarily, and an in-call-modification procedure in which the characteristics of the speech circuit can be modified during the call. Additional facilities are also provided to allow different versions of ISUP to interwork.

The ISUP provides a greater range of supplementary services than that adopted for the TUP. The means of implementing the supplementary services is more flexible and a wider range of customer features is provided. The ISUP provides a user-to-user signalling capability in which customers can exchange data over the signalling channel without it being analysed by the network.

6.10 References

1 CCITT Recommendations Q.721-Q.725: 'Telephone user part' (ITU, Geneva)
2 CCITT Recommendations Q.761-Q.766: 'Integrated Services Digital Network User Part (ISDN UP)' (ITU, Geneva)
3 CCITT Recommendations Q.741 and X.61: 'Data User Part' (ITU, Geneva)
4 CCITT Recommendation Q.723: 'Formats and codes' TUP (ITU, Geneva)
5 CCITT Recommendation Q.724: 'Signalling procedures' TUP (ITU, Geneva)
6 CCITT Recommendation Q.763: 'Formats and codes' ISUP (ITU, Geneva)
7 CCITT Recommendation Q.764: 'Signalling procedures' ISUP (ITU, Geneva)
8 CCITT Recommendation Q.730: 'ISDN supplementary services' (ITU, Geneva)

Chapter 7

Transaction capabilities

7.1 Introduction

Transaction capabilities (TC) is a protocol that is used, in conjunction with an appropriate network-layer service (e.g. the SCCP and MTP), to provide the non-circuit-related transfer of information across networks. TC is defined by CCITT [1] as part of CCITT No. 7. However, because it is defined in accordance with the OSI 7-layer model (see Chapter 4), TC is a general protocol that can be applied in a broad range of applications. These applications include:

(*a*)   Mobile-service location registration
(*b*)   Access to specialised network nodes
(*c*)   Provision of advanced supplementary services
(*d*)   Operations, maintenance and network management

In many forms of mobile service (e.g. car telephones in a land-mobile service), the location of the mobile (the car) is regularly fed into 'location registers'. These allow the network to determine the location of the car at any time and thus assist in the provision of incoming calls to the car. The process of updating the registers with the location of the car involves the non-circuit-related transfer of information across the network. TC can be used to control the transfer of the location information.

It is sometimes appropriate to store information in specialised nodes within the network. For example, if specialised routeing information is specific to only one service, it might not be convenient to store the routeing information in a large number of exchanges in the network. In this case, a network database can be used to store the information. When an exchange needs to gain access to the routeing information, a non-circuit-related information transfer occurs between the exchange and the database. This transfer can be enacted by TC.

Advanced supplementary services can employ TC to provide non-circuit-related information transfer. For example, a 'look-ahead' procedure can be adopted. In this facility, a local exchange originating a call can send a non-circuit-related message to the destination local exchange, checking whether or not the called customer is free. If the called customer is free, the local exchange can proceed with the establishment of a traffic path through the network. If the called customer is busy, then the local exchange can abort the call without having established a traffic path. Thus the use of traffic paths can be made more efficient by only setting-up calls for which the called customer is free.

Modern networks require a comprehensive operations, maintenance and network management infrastructure to ensure reliable service. These functions can require the transfer of large amounts of data between nodes (between exchanges, between exchanges and operations centres and between operations centres) or they can simply require the transfer of an instruction and a corresponding response. TC can be used to control the transfer of this information because of its non-circuit-related nature. The operations, maintenance and administration part (OMAP) of CCITT No. 7 uses TC to monitor and control the signalling network. Further details are given in Section 7.5.

The applications of TC can be categorised into those requiring 'real-time' responses and those requiring 'off-line' responses. In real-time applications, information transfer is required quickly to complete a function. For example, if a local exchange needs to gain access to a network database for specialised routeing information during the establishment of a call, every second counts. The time taken to achieve the information transfer adds to the post-dialling delay encountered by the calling customer. Another feature of real-time applications is that they generally involve the transfer of a small amount of data. In the example, a telephone number is transmitted to the database and routeing information is returned: this requires little data.

In off-line applications of TC, the length of time taken to achieve an information transfer is not critical. For example, if there is a need to transfer bulk statistical data from an exchange to a central maintenance centre, the timing in seconds, or even minutes, is not a critical factor. More important in this case is the need for secure transfer of the information. The example demonstrates another feature of off-line applications: in general, they involve the transfer of large amounts of data.

Section 5.6.2 describes the services offered by the network layer of the OSI model as being either connectionless or connection-oriented. Real-time applications use the connectionless services offered by the network layer, whereas off-line applications use the connection-oriented services. To make early progress, CCITT [2] has concentrated on TC based on the connection-less-network service. TC based on a connection-oriented-network service is not yet specified. Hence, whilst off-line applications will be able to use TC in the future, only real-time applications can currently be handled. This is sensible, since the need for real-time applications (e.g. database access) is more pressing than for off-line applications.

An objective of TC is that it should be application-independent. This means that TC should be a general protocol that is flexible enough to handle any non-circuit-related application, including those not yet specified. In this sense, TC is a unique element within CCITT No. 7. Its counterparts (the ISUP, TUP and DUP) are specifically designed for certain applications, e.g. the TUP's prime function is to control the set-up and release of telephone calls. The reason for the application-dependence of user parts is that both the information-transfer function (signalling system) and the exchange-call-control function are directly concerned with setting-up and releasing circuit-switched calls. It is therefore implicit that there is a a very close relationship between the two functions. For non-circuit-related applications, the aim is to break the relationship between the control function and the information-transfer function. This is achieved by specifying the two functions independently, TC being defined as a flexible information-transfer mechanism. To help this approach, TC defines elements of information termed 'components'. A component is used to request a remote node to perform an action (termed an 'operation'). A component is also used to return the results of an action. The format of components is standardised as part of TC and is defined in an application-independent manner. Included within each component requesting an action to be performed is an operation code that is application-dependent and is selected by the application. Hence, an application can draw upon components to achieve information transfer required for that particular application, the application-dependent part being restricted to particular fields. Further details are given in Section 7.2.

TC can use the SCCP and MTP to transfer messages across the network. However, for maximum flexibility and applicability, TC is designed to conform with the OSI 7-layer model. Hence, in the future, TC should be able to use any transfer mechanism that provides an OSI network service. For example, if a network uses a packet-data transfer mechanism [3] as part of its existing operations and maintenance infrastructure, TC can be used to control appropriate information flow without modifying the transfer mechanism. This demonstrates the powerful nature of the OSI-model approach and it is a step in the direction of allowing network operators to 'mix and match' protocols according to local circumstances.

TC can be described in terms of its architecture (outlined in Section 7.2), formatting principles (Section 7.3) and procedures (Section 7.4). The OMAP, as a user of TC, is described in Section 7.5.

7.2 Architecture

In the customer-to-customer context described in Section 4.5, the functions of TC are usually regarded as belonging to the network layer (Layer 3) of the OSI 7-layer model. However, on a node-to-node basis, the full 7 layers can be applied to non-circuit-related communications. In this context, the combination of the MTP and SCCP provides the functions of Layers 1 to 3 of the model. Layers 4 to 7 of the model are covered by TC and the TC user, as shown in Fig. 7.1. TC consists of an intermediate-service part (ISP) and a transaction-capabilities application part (TCAP).

The ISP covers the functions of Layers 4 to 6 of the OSI model, thus handling the transport, session and presentation aspects of non-circuit-related communication. The ISP is required when TC is based on a connection-oriented-network service. As explained earlier, the use of TC based on a connection-oriented-network service is not yet defined. Hence, details of how Layers 4 to 6 are applied to TC are not yet specified. When the connectionless-network services are used by TC (real-time applications), the functions of ISP are not required and Layers 4 to 6 can be considered to be transparent.

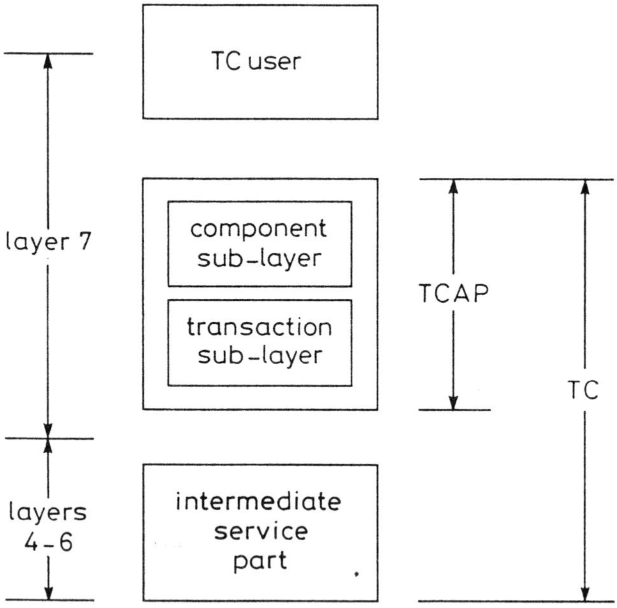

Fig. 7.1 Structure of transaction capabilities

The functions of Layer 7 of the OSI model are covered by TCAP and the TC User. One aim of TC is that it should be application-independent. This is achieved by defining the application-independent functions within TCAP and the application-dependent functions within the TC user. The TC user, which is not specified as part of TC or CCITT No. 7, thus includes the functions specific to a particular application. For example, in the case of a network database storing specialised routeing information, TCAP (in conjunction with the SCCP and MTP) is responsible for transferring non-circuit-related information between exchanges in the network and the database. The formats and procedures used in TCAP are common to those used for all other applications. In the example, the TC user is the control-function of the database: this is responsible for analysing the messages delivered by TCAP and returning responses. Thus, the TC user provides the application-specific functions.

TCAP consists of the transaction sub-layer and the component sub-layer, as illustrated in Figure 7.1. The transaction sub-layer is responsible for establishing and maintaining a connection (transaction) between two nodes. The component sub-layer is responsible for initiating actions at a remote node and returning the results of such actions.

The component sub-layer in one node communicates with the corresponding sub-layer in another node by sending and receiving 'components'. A component consists of a request to perform an operation (or an action) or a response to a request. For example, consider an exchange that has received a telephone number from a calling customer that needs to be translated into specialised routeing information by a network database. The exchange sends a component to the database requesting that a translation be performed. A parameter in the component gives the telephone number. Upon completion of the translation by the database, a component is returned to the exchange, responding to the request. The responding component contains a parameter including the specialised routeing information.

The operations that are requested by the component sub-layer can be divided into four categories, termed classes, corresponding to the level of response expected upon completion of the operation. In Class 1, both success and failure to conduct an action are reported back to the node initiating the request for an operation to be performed. An example of this type of operation is when an exchange requests a remote database to perform a routeing translation on a telephone number. It is important in this case for the database to report back to the exchange either the successful completion of the operation (with the result being a translated number) or the failure to complete the operation (with a reason for the failure).

In Class 2, only failures to complete the operation are reported. This category can be used when, for example, there is a need to conduct a routine test and a reply is only necessary when there is a fault preventing the completion of the test.

Class 3 operations are used when it is necessary to report only successful results. This can be used in the case when a fault is suspected and the likely outcome is the failure of the operation. It is assumed that the operation fails unless a successful result is reported back.

If neither success nor failure needs to be reported, then Class 4 can be used. An example of its use is when a node wishes to send a warning of an event to several other nodes and a response or acknowledgement is not required.

Successive components exchanged between two TC users constitute a 'dialogue'. Responsibility for the dialogue rests with the transaction portion. In an 'unstructured dialogue', TC users send components without an explicit association being established. This can be likened to the connection-less service described in Section 5.6.2 for the SCCP. An unstructured dialogue is typically used for components for which a reply is not expected.

In a 'structured dialogue', an explicit association is formed between two nodes. This can be likened to the connection-oriented service described in Section 5.6.2 for the SCCP. The dialogue has three phases (begin, continue and end) similar to the establishment, conversation/data and release phases described for circuit-switched calls. The structured form of dialogue is typically used for components which are part of a series of interchanges between nodes.

7.3 Format principles of TCAP

7.3.1 General message format

TCAP messages consist of 'information elements', each having a standard structure. An information element consists of three fields and they are always in the same order. The fields correspond to the name/length/ information structure described for the SCCP and ISUP in Sections 5.6.3 and 6.7 respectively. For TCAP, CCITT [4] uses the terminology tag/length/ contents and this terminology will be used in this chapter. Fig. 7.2 illustrates the structure of an information element. The tag distinguishes between one type of information element and another and governs the meaning of the contents field. The length field indicates the length of the contents field. The contents field contains the substance of the information element, i.e. the primary information that the element conveys. The contents can consist of a single value or one or more information elements. If the contents consist of a single value, CCITT [4] calls the information element a 'primitive'. Care must be taken to avoid confusion with the primitives defined for use in the OSI model, which apply to inter-layer communication. A primitive-information element is shown in Figure 7.3. If the contents consist of one or more information elements, then the encompassing information element is called a 'constructor', as shown in Figure 7.4 This recursive approach allows the formatting of TCAP to be extremely flexible whilst remaining application-independent. Each application can make use of primitive or constructor information elements to build simple or complex messages meeting the needs of that application.

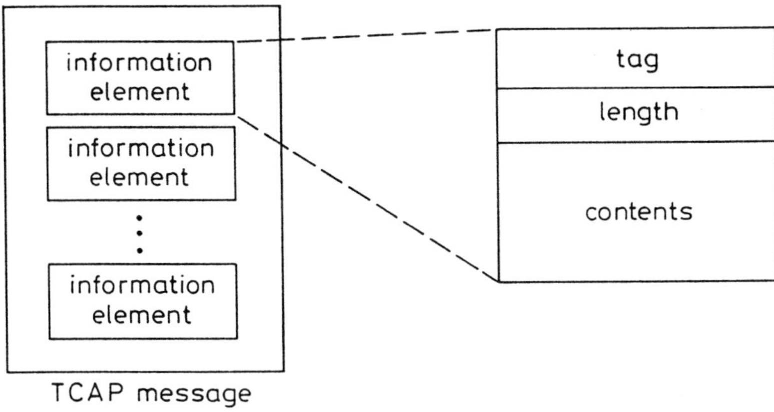

Fig. 7.2 Structure of information element

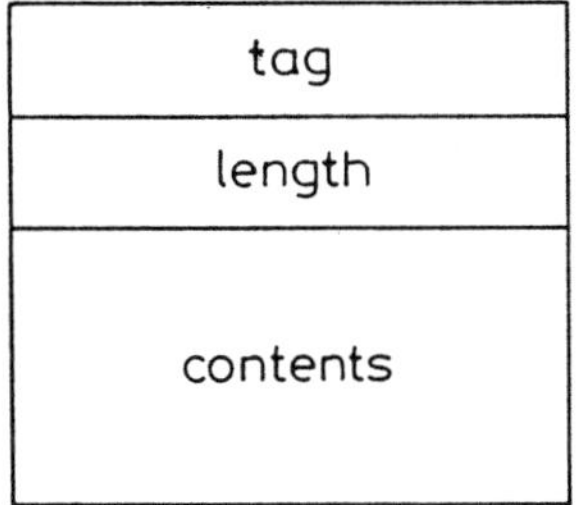

Fig. 7.3 Primitive information element
Source: CCITT Recommendation Q. 773

A TCAP message consists of a constructor information element. The message contains a 'transaction portion', containing information elements pertinent to the transaction sub-layer, and a 'component portion', containing information elements pertinent to the component sub-layer. The format is illustrated in Fig. 7.5.

The message commences with a message-type tag, indicating whether the message refers to an unstructured dialogue or a structured dialogue. In the case of structured dialogue, the message-type tag indicates the phase of the dialogue, i.e. whether the message refers to the begin, continue or end phase. The message-length field indicates the length of the contents field. The contents field consists of a series of transaction-portion information elements (TPIEs), each following the general format of tag/length/contents. The TPIEs pertinent to the transaction sub-layer form the transaction portion. One TPIE contains the component portion and consists of the component-portion tag, component-portion length and component-portion contents. In an extension of the recursive approach, the component-portion contents consist of a series of component-information elements, each preceded by a component-type tag and component-length field.

The recursive approach used by TCAP, in which the contents field of one information element contains the tag/length/contents of other information elements, is an important difference between the formatting technique of TCAP and that of the ISUP. Whereas the ISUP technique is flexible, it is not application-independent. TCAP requires an application-independent approach. The recursive use of tag/length/contents can increase the processing overhead of a message: for example, in simple messages, some of the information that is implicit in the message type has to be provided explicitly to conform with the overall message structure. However, the technique is extremely flexible and this far outweighs the disadvantages of the tag/length/contents approach for non-circuit-related applications.

7.3.2 Transaction portion

In the case of unstructured dialogue, an explicit association between two nodes is not established by the transaction sub-layer. The message type used in an unstructured dialogue is termed 'unidirectional', reflecting the feature

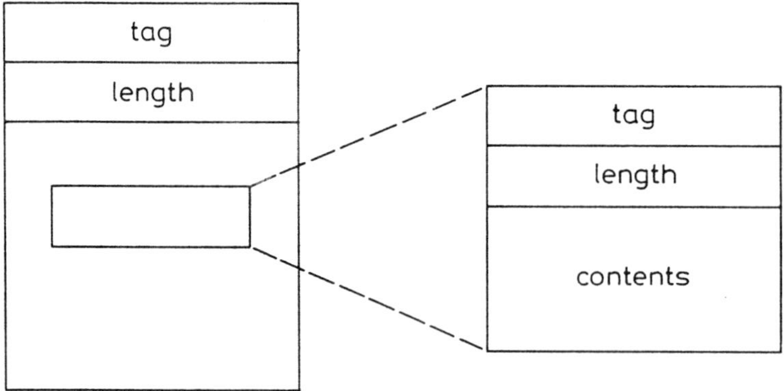

Fig. 7.4   Constructor information element
Source: CCITT Recommendation Q. 773

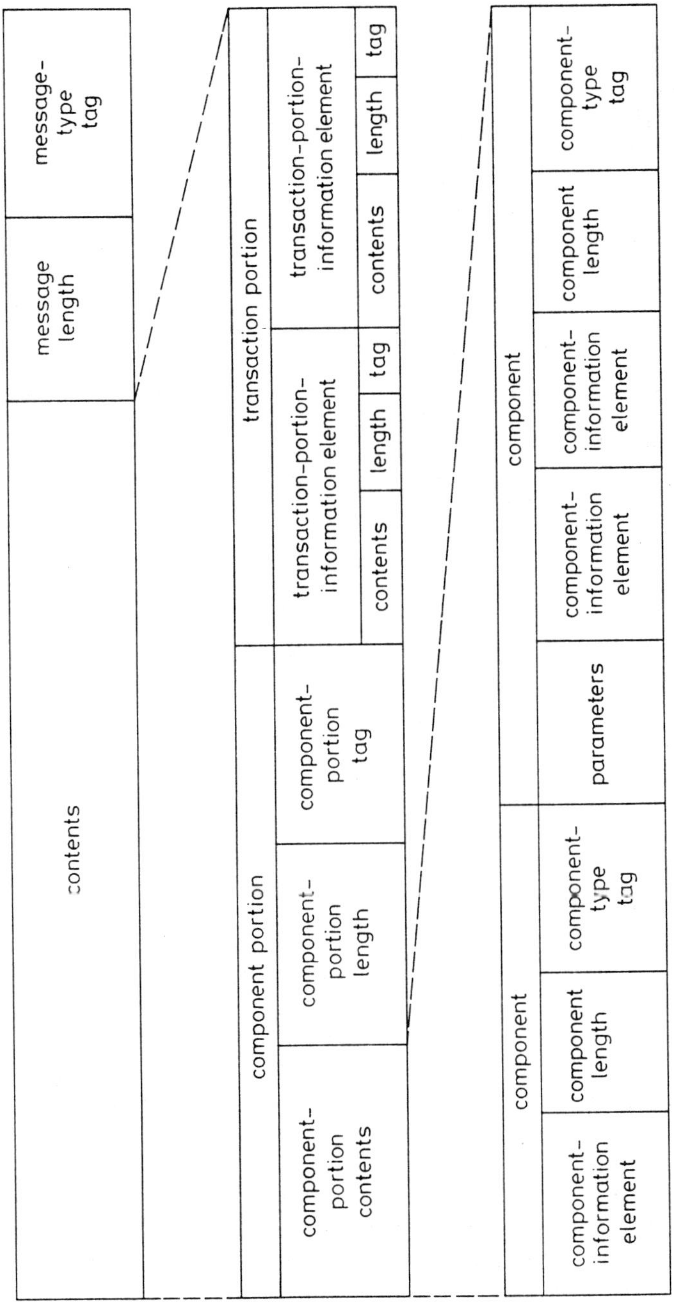

Fig. 7.5   Principle of TCAP message format

component– portion contents	component– portion length	component– portion tag	message length	message- type tag

Fig. 7.6   Format of unidirectional message type

that an answer is not normally expected. The format of the message primarily comprises the component portion, as illustrated in Fig. 7.6.

In the case of structured dialogue, the transaction sub-layer establishes an explicit association between two nodes. This is achieved using three message types: 'begin', 'continue' and 'end'. The begin message is illustrated in Fig. 7.7. The dialogue is deemed to commence by including an 'originating-transaction-identity' consisting of a tag, length and contents. A transaction identity is a number generated by a node to identify to which dialogue a message refers. The originating-transaction identity is a number selected by the node generating the begin message.

component– portion contents	component– portion length	component– portion tag	OTID	OTID length	OTID tag	message length	message- type tag

OTID originating transaction identity

Fig. 7.7   Format of begin message

The continue message type is illustrated in Fig. 7.8. One or more continue messages can be exchanged between nodes as part of the structured dialogue. The continue type of message includes both an originating-transaction-identity and an destination-transaction-identity information element. Thus, both the originating and destination nodes can associate a continue message with a particular dialogue. It is the inclusion of transaction identities that maintains the association and thus allows a structured dialogue to take place. The originating identity always refers to the node generating the particular message. Details of the method of coding the identity fields are given in Section 7.4.

component-portion contents	component-portion length	component-portion tag	DTID	DTID length	DTID tag	OTID	OTID length	OTID tag	message length	message-type tag

OTID originating transaction identity
DTID destination transaction identity

Fig. 7.8   Format of continue message

When either node involved in a dialogue has completed the transfer of information relevant to that transaction, the dialogue is ended by the transmission of an end message. The format of the end message is illustrated in Fig. 7.9.

component-portion contents	component-portion length	component-portion tag	DTID	DTID length	DTID tag	message length	message-type tag

DTID destination transaction identity

Fig. 7.9   Format of end message

CCITT [4] also defines two 'abort' message types. These are used when the transaction sub-layer or TC user cannot process a message due to an error. For example, if an unrecognised message type is received, an abort message is returned with an explanation of the fault.

7.3.3 Component portion

The component portion contains one or more components, each component being a constructor information element. The component portion uses three basic types of component to initiate actions and return the results of actions: 'invoke', 'return-result (last)' and 'return-result (not last)'. The return-result (not last) component is used when there is too much data to be included in a return-result (last) component. Two other components, 'return error' and 'reject', are used to indicate that a request to carry out an action has not been completed or that aspects of a message received (e.g. a component) are unrecognised.

The invoke component is used to request a node to perform an action. For example, one node might request another node to perform a routeing translation: an invoke component is used for this purpose. An example of the format of the invoke component is shown in Fig. 7.10. The component-type tag indicates that the component is an invoke type. The invoke-identity information element contains a reference number that uniquely identifies an operation within a dialogue. This allows the originating node to correlate any response with the initial request to perform an action. The operation-code-information element indicates the precise operation (or action) that is to be performed (e.g. perform a translation). To maintain the application-independence of TCAP, the operation-code-information element is determined by the application and is thus not specified as part of TCAP. The component portion does not interpret or analyse the meaning of the operation code. The code is received from the TC user, inserted into the appropriate location in the component portion, transmitted to the destination node and delivered to the destination TC user. It is the TC users that determine the meaning of the operation-code-information element. Parameters may be included in the message if appropriate; e.g. if the operation requested is to perform a routeing translation upon a telephone number, the number can be included as a parameter following the operation-code-information element.

optional parameters	op. code	op. code length	op. code tag	invoke identity	invoke identity length	invoke identity tag	comp. length	comp. type tag

comp.   component
op.   operation

Fig. 7.10   Example of an invoke component

The return-result (last) component is used to report the successful completion of an operation. The return-result (not last) component is used to indicate that only part of the results of an operation is being supplied, with the implication that the remainder of the results will be supplied. To complete the reporting of results in this case, a return-result (last) component is used.

The return-result (last) and return-result (not last) components have the same general format, similar to that for the invoke component shown in Fig. 7.10. The invoke-identity-information element is included to correlate the result of the operation with the original invocation. The unique reference number allows the node initiating the invocation of an action to understand to which invocation the results pertain. The components can include optional parameters. For example, if the original invocation requested a translation of a telephone number, the parameters field in the return-results component may contain the translated number. If parameters are present, then the operation-code field is also required. If parameters are not present, the operation-code field can be omitted.

7.4 TCAP procedures

7.4.1 General

The procedures for TCAP are restricted to facilitating the exchange of components between TC users. This approach supports the application-independence of TC. Any procedures in addition to those specified for TCAP that are necessary to implement a service are specified as part of the application. TCAP procedures are divided into component sub-layer procedures and transaction sub-layer procedures.

7.4.2 Component sub-layer procedures

Component sub-layer procedures are very flexible to ensure that they can handle the requirements of a range of applications. Examples of component flows are given in Figs. 7.11 to 7.13 to illustrate the principles.

Fig. 7.11 shows the basic component flow in which Node A sends an invoke component to Node B and Node B returns the results to Node A. An example is the 'look-ahead' procedure described earlier. This feature is not yet specified internationally, but the following describes a typical sequence. Suppose that Node A is a local exchange that needs to establish a call to another local exchange (Node B). In this case, Exchange A sends an invoke component to Exchange B requesting Exchange B to determine whether or not the called customer is free. For this example, the operation-code-information element is 'check customer status' and a parameter is included in the component containing the called customer's number. After checking the status of the called customer, Exchange B indicates whether or not the customer is free by returning a return-result (last) component. This information allows Exchange A to determine whether or not to establish a traffic path through the network.

In Fig. 7.12, Node A sends an invoke component (1) to Node B, but Node B needs more information before processing the component. In this case, Node B initiates its own invoke component (2), soliciting a response from Node A in a return- result (last) component (2) . Upon analysis of the result, Node B responds to the original invocation with a return-result-(last)

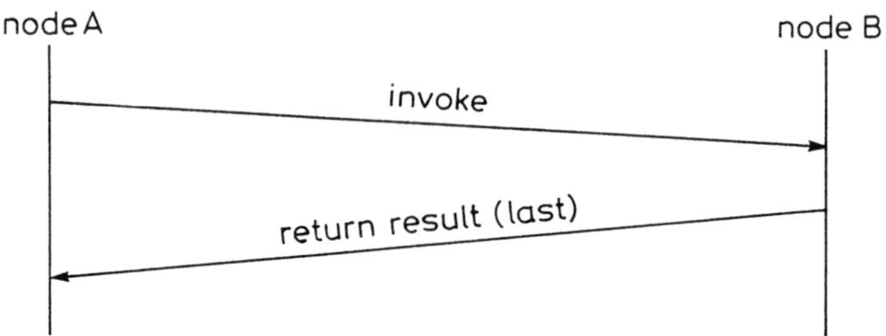

Fig. 7.11   Example of basic component flow

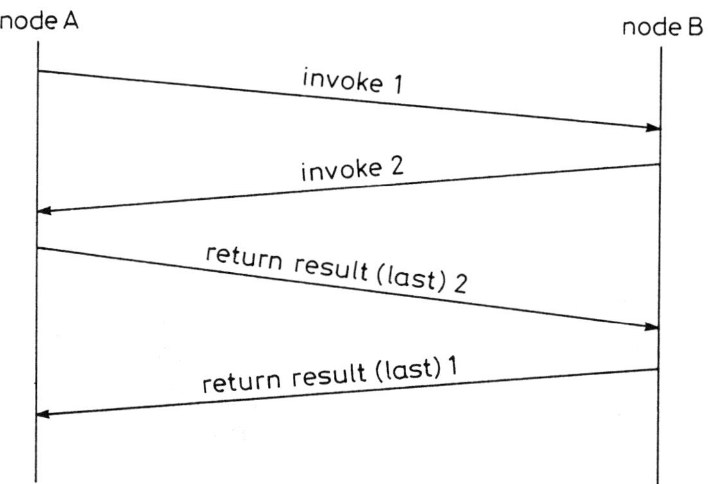

Fig. 7.12.   Example of multiple invocation

component (1). An example of this sequence is when Node A is an exchange seeking the translation of a telephone number into routeing information from a database (Node B). In the example, the database needs more information from Exchange A, e.g. the calling customer's number might be required to provide appropriate routeing information. Once this information is provided to the database, the original invocation can be processed and the routeing information is supplied to Exchange A in a parameter within a return-result (last) component.

In Fig. 7.13, the information resulting from processing an invoke component at Node B is too long to be sent in a single return-result component. Hence, the information is segmented into the parameters of three components, two of the components being of the type return-result (not last) and the final component being of the type return-result (last).

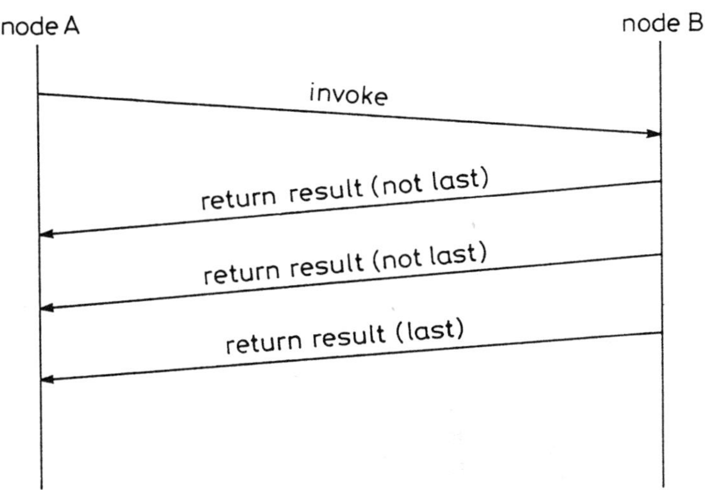

Fig. 7.13   Example of segmented result

The CCITT specification of TCAP procedures [5] includes a whole range of procedures to deal with abnormal conditions; e.g. if an invoke component is received with a syntax error, a reject component is returned indicating the reason for the failure. Details of these abnormal conditions are not given here.

7.4.3 Transaction sub-layer procedures

In the unstructured dialogue case, the transaction portion of a node sends a unidirectional message to another node, as illustrated in Fig. 7.14. Upon receipt of a unidirectional message, the transaction sub-layer in Node B passes the information to the TC user at Node B without analysis.

An example of the transaction sub-layer procedures in a structured dialogue is given in Fig. 7.15. Node A initiates the commencement of a structured dialogue by sending a begin message. The originating-transaction identity (OTID) selected by Node A, and included in the begin message,

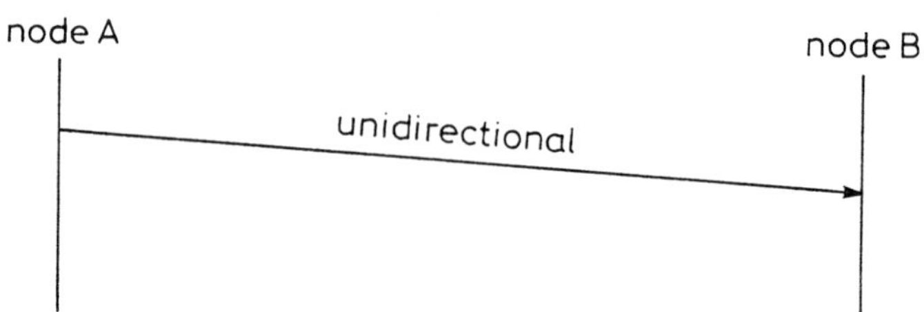

Fig. 7.14   Transaction sub-layer procedure for unstructured dialogue

is X. Node B analyses the begin message and agrees to establish a dialogue. Node B returns a continue message to confirm this decision. Node B selects the OTID of Y for inclusion in the continue message. The destination-transaction-identity (DTID) field contains the identity X (corresponding to the number selected by Node A).

Upon receipt of the continue message from Node B, Node A analyses the information and sends a continue message to Node B. In this case, the OTID is X and the DTID is Y. After reception and analysis of the continue message from Node A, Node B determines that the dialogue can now be terminated and returns an end message. There is no OTID in the end message and the DTID is X.

In the example above, Node B initiated the end of the dialogue, but it would be just as appropriate for Node A to perform this function, depending upon the application. The case when either node initiates an end message is termed the 'basic' method of ending the dialogue. There is another method of ending a dialogue termed the 'pre-arranged' end. A typical application of the pre-arranged end is when a node needs information from a database but it does not know which database to query. In this case, a request is broadcast to numerous databases with the anticipation that only one database will respond positively. To avoid all databases responding (all but one responding negatively), the dialogue is deemed to have ended unless a positive response is received. The dialogue between the node and the database responding positively then continues as described above.

7.5 Operations, maintenance and administration part (OMAP)

7.5.1 General
As mentioned earlier, a user of TC is the operations, maintenance and

administration part (OMAP) of CCITT No. 7. OMAP allows operations and maintenance personnel to monitor and control the equipment associated with the CCITT No. 7 signalling network. Thus, operations and maintenance personnel can manage the signalling network from a central location; CCITT No. 7 itself provides the means to communicate with all other nodes in the network. TC is used to provide a non-circuit-related transfer of information between the controlling point and the node(s) involved with providing specific functions.

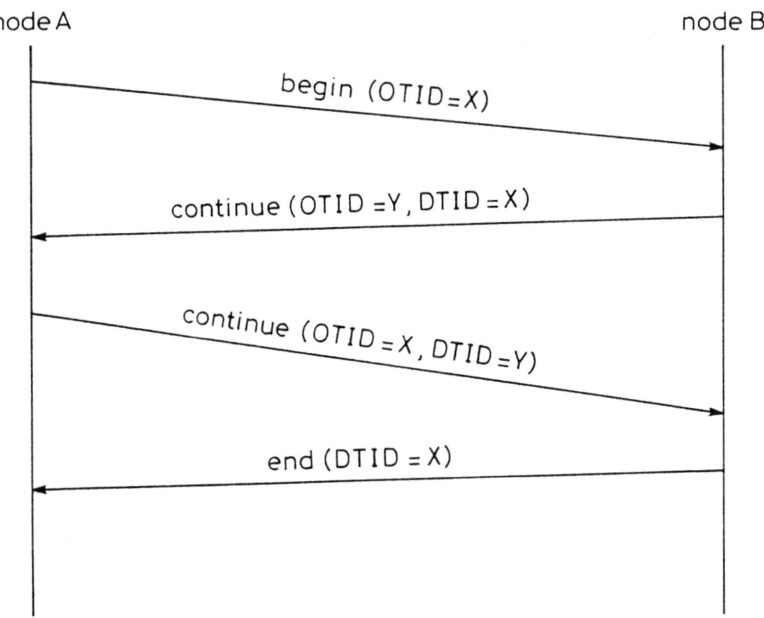

Fig. 7.15   Transaction sub-layer procedures for structured dialogue

In terms of architecture, the combination of OMAP and TCAP forms Layer 7 of the OSI model, in the node-to-node context described in Chapter 4. In addition, OMAP provides some of the functions of the application that lies above Layer 7. The connection-oriented service for TC is not yet specified. Hence, the specified OMAP features are defined to make use of the TC connectionless service.

The functions provided by OMAP [6] can be summarised to be the management of routeing data, the circuit-validation test, the MTP-routeing-verification test and measurement reporting. Many of the details are not yet specified (e.g. many message formats), but a brief description is given below to illustrate the use of TC.

7.5.2 Management of routeing data

Each signalling point in the network holds routeing data that are used to route messages from one node to another. For an efficient signalling network, it is essential that operations and maintenance personnel can remotely monitor and control such data. Thus, management procedures are being defined to use TCAP to add, modify or delete routeing data contained at remote-signalling points. Procedures are also defined to interrogate the data held in a particular node (e.g. to check that such information is not corrupted).

7.5.3 Circuit-validation test (CVT)

Consider two exchanges connected by traffic circuits. Each exchange holds translation data that are used to place a call on a specific traffic circuit. The CVT procedure allows operations and maintenance personnel to ensure that the two exchanges hold consistent data that allow a call to be made. The procedure can be used if, for example, a fault prevents the use of a particular traffic circuit. If the performance of the signalling system related to the traffic circuit can be proved to work, then the fault can be attributed to the control of the traffic circuit itself.

The procedure works by stimulating the generation of a circuit-identity code (CIC) at each exchange, i.e. at each end of the traffic circuit. The two values are compared and if they are the same, then the signalling data used for the traffic circuit can be assumed to be correct. If the two values are not the same, then signalling data at one of the exchanges can be assumed to be corrupted and further investigation can be undertaken.

7.5.4 MTP-routeing-verification test (MRVT)

Each exchange in a CCITT No. 7 signalling network holds data that are used by the MTP to route messages. This data can be very complex, especially if numerous signal-transfer points are used. The purpose of the MRVT is to ensure that the data throughout the network is consistent. For example, the test verifies that messages are not being routed in a continuous 'loop' (i.e. never reaching a destination, but being routed from one exchange to another). The test also verifies that if one signalling point can send a message to another, then the reverse routeing also applies. The MRVT is initiated whenever new MTP data are introduced (or existing data changed) or at the request of operations and maintenance personnel.

The procedure involves a signalling point sending a 'MTP-Routeing-Verification Test' message to a specified-destination signalling point. As the message is routed through the network, a list of the signal-transfer points used is recorded. When the message arrives at the destination-signalling point, the list of STPs is returned to the initiator of the procedure to check the data against stored records.

7.5.5 Measurement reporting

To manage the signalling network effectively, it is necessary to measure the performance and availability of appropriate equipment. Procedures are defined to initiate and stop the measurements being made. The measurements can be made and transferred on a regular basis (e.g. for general management of the network) or on demand (e.g. during an investigation into network efficiency or fault conditions).

7.6 Chapter summary

Transaction Capabilities (TC) is a protocol that, in conjunction with an appropriate network-layer service (e.g. the MTP and SCCP), provides non-circuit-related transfer of information between nodes. TC conforms to the OSI 7-layer model and it is therefore a general protocol that can be used by a wide range of applications, e.g. the operations, maintenance and administration part (OMAP) of CCITT No. 7. TC is designed to be application-independent. TC will eventually be defined for use in real-time and off-line circumstances. However, only real-time aspects of TC are currently defined, making use of a connectionless network service.

TC lies within Layers 4 to 7 of the OSI 7-layer model. Layers 4 to 6 are transparent for connectionless-network services. The transaction-capabilities application part (TCAP), together with some functions of the TC user, comprises Layer 7 of the OSI model.

TCAP consists of the transaction sub-layer and the component sub-layer. The transaction sub-layer is responsible for establishing and releasing TC connections. The component sub-layer is used to request actions to be performed and to report the results of actions.

TCAP messages consist of information elements with the general format of tag/length/contents. The contents can consist of a single value. However, in general, the contents of one information element comprise another information element in a recursive approach to message structure. Such information elements are termed constructors.

The transaction sub-layer defines three message types: begin, continue and end. The begin message initiates a transaction, continue messages are used during the transaction and the end message releases the transaction.

The component sub-layer defines several components. The invoke component is used to request an action to be performed by another node. The response can be successful (in which case, a return-result component is returned) or unsuccessful (in which case, a return-error component is returned). The return-result component can be qualified as last or not-last. A reject component is also used in failure conditions.

The operations, maintenance and administration application part (OMAP) is a user of TC. It allows operations and maintenance personnel to manage the signalling network from central locations. Part of OMAP is within Layer 7 and part is within the application. In the OMAP, management of routeing data allows the monitoring and control of routeing data in nodes. Circuit-validation testing is used to confirm the consistency of traffic-circuit data. MTP-routeing-verification testing allows the routeing capability of the MTP to be checked. Measurement reporting provides the capability to gather measurements from various parts of the signalling network.

7.7 References

1 CCITT Recommendations Q.771-775, 'Transaction capabilities' (ITU, Geneva)
2 CCITT Recommendation Q.771: 'Functional description of transaction capabilities' (ITU, Geneva)
3 CCITT Recommendations X.25 and X.75: 'Packet mode data' (ITU, Geneva)
4 CCITT Recommendation Q.773: 'TCAP Formats' (ITU, Geneva)
5 CCITT Recommendation Q.774: 'TCAP Procedures' (ITU, Geneva)
6 CCITT Recommendation Q.795: 'Operations, maintenance and administration part' (ITU, Geneva)

Chapter 8

DSS1 Physical and data-link layers

8.1 Introduction

Customers need a flexible communications system that can provide increasingly-sophisticated services. In response to this need, network operators throughout the world are implementing integrated services digital networks (ISDNs) that provide customer-to-customer connections using digital transmission links and software-controlled exchanges. ISDNs need responsive and efficient signalling systems for use within the network and for use between customers and the network. In the past, international-standards organisations (e.g. CCITT) have concentrated on the specification of inter-nodal signalling systems. However, the benefits of ISDNs can only be realised by customers if the access-signalling systems are similarly responsive. Flexible information transfer is required between customers and the network and directly between customers. The digital subscriber signalling system No. 1 (DSS1)[1] is defined to meet the demanding requirement of providing flexible signalling for customers and is a step towards the objective of providing an unimpeded signalling-transfer capability.

DSS1 is responsible for transferring information between customers and local exchanges in both circuit-switched and packet-data applications. Chapter 4 describes the architecture of DSS1, and in particular, its conformance with the OSI 7-layer model. This chapter describes the functions of the physical and data-link layers of DSS1. Chapter 9 describes the functions of the network layer.

8.2 Physical layer

Layer 1 describes the physical, electrical and functional characteristics of the interface between the customer and the local exchange (network). Access from a customer to a local exchange in an ISDN is by one of two connection types: basic access and primary (multi-line) access.

The basic access is shown in Fig. 8.1. The transmission link between the customer (user) and the local exchange provides an information-bit rate of 144 kbit/s. The bit rate is structured to allow two traffic channels at 64 kbit/s each (B channels) and a signalling channel at 16 kbit/s (D channel). Although the D channel is dedicated to controlling the two B channels, DSS1 is still a common-channel-signalling system because signalling capacity is allocated dynamically according to need. The signalling channel can handle numerous terminals at the customer's premises by dynamically allocating signalling capacity as required.

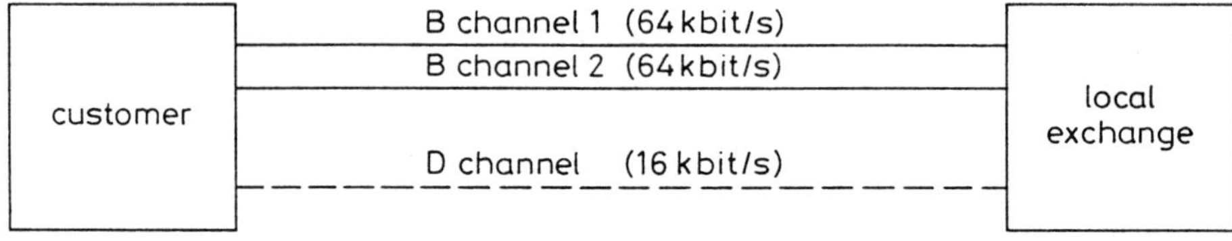

Fig. 8.1   Basic access

The primary (multi-line) rate access is shown in Fig. 8.2. This access technique uses a transmission link in which a 64 kbit/s signalling channel (D channel) controls up to 30 traffic channels (B channels).

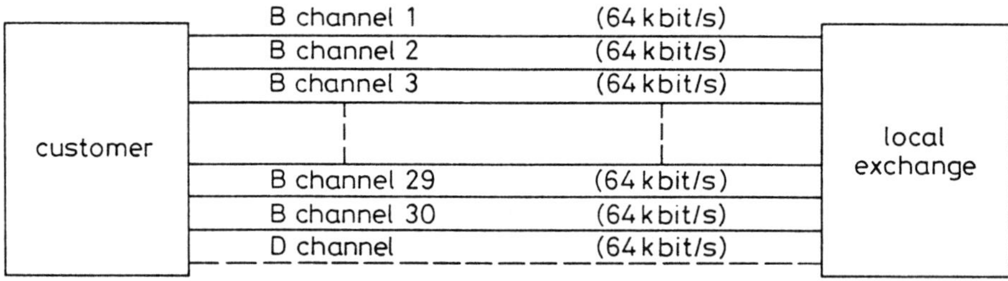

Fig. 8.2   Primary (multi-line) rate access

The characteristics of Layer 1 functions allow the ability to:

(*a*)   Connect a number of customer terminals to a line, each terminal having access to the B channels for traffic and the D channel for signalling

(*b*)   Supply power from the exchange for basic telephony in the case of the failure of customer-provided power

(*c*)   Provide point-to-point and broadcast working, as described in Section 4.8.

8.3 Data-link layer functions

Layer 2 (data-link layer) is responsible for transferring messages between a customer and a local exchange. The messages carry information (generated by Layer 3) in an information field. The Layer-2 functions do not understand the meaning of the information field, but it is their job to deliver the information with a minimum of loss or corruption. The functions of Layer 2 can be described in terms of the formats used and procedures adopted [2], as described in Sections 8.4 and 8.5 respectively. The formats and procedures are based on a high-level-data-link control (HDLC) protocol [3] defined by the International-Standards Organisation.

8.4 Layer 2 formats

The exchange of information by the data-link layer takes place in messages, termed 'frames', which are similar to the signal units described for CCITT No. 7 in Section 5.4.1. The frames are designated as being either 'commands' (requesting an action to be performed) or 'responses' (reporting the result of a command). The general format of a frame is given in Fig. 8.3.

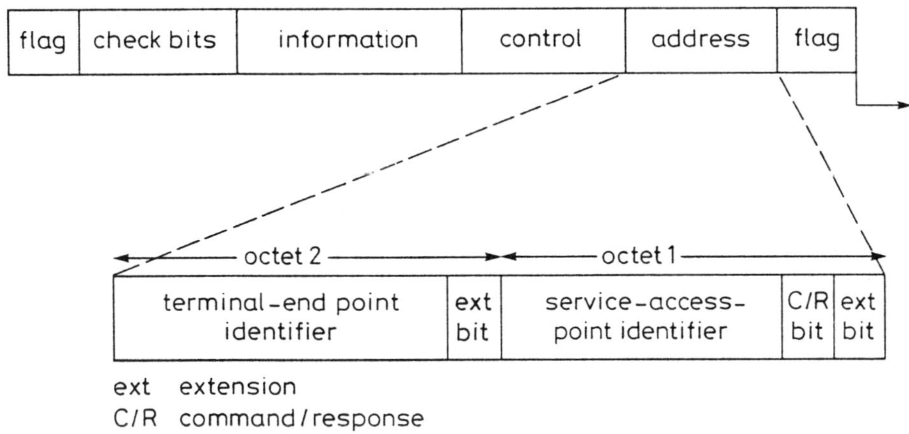

Fig. 8.3   Format of a frame

Each frame starts and ends with a flag, which is a unique code that delimits the frame. The flag is coded 01111110. Imitation of the flag by any other field in the frame is avoided by banning any consecutive sequence of more than five ones. This is achieved by a data-link-layer function that inserts a zero after any sequence of five consecutive ones, except in the flag, before transmitting a frame. Upon receipt of a frame, any zero that is detected following five consecutive ones is discarded.

The address field consists of two octets. The command/response (C/R) bit of the address field indicates whether the frame is designated as a command or a response. If the frame is designated as a command, the address field identifies the receiver (i.e. the local exchange or customer equipment receiving the frame). If the frame is designated as a response, the address field identifies the sender (i.e. the local exchange or customer equipment sending the frame). The extension bits illustrate a technique for expanding the length of a field in a flexible manner. The extension bit in the first octet is set to value 0 to indicate that another octet follows. The extension bit in the second octet is set to value 1 to indicate that the second octet is the last of the address field. If, in the future, it is necessary to expand the size of the address field, the extension bit in the second octet can be changed to value 0 to indicate that a further octet follows. The third octet, in this case, would have an extension bit of value 1 to indicate that it is the last octet. The increase in size is achieved without affecting the rest of the frame.

The service-access point identifier (SAPI) and terminal-endpoint identifier (TEI) are used to identify the connection and terminal to which a frame pertains. The service-access point is used to describe the location at which the data-link layer provides services to Layer 3. Thus, the SAPI indicates which entity at Layer 3 needs to analyse the contents of the information field. For example, the SAPI indicates that the contents of the information field are relevant to the call-control procedures for circuit-switched calls or to the call-control procedures for packet data. The (TEI) indicates the terminal equipment to which the message refers. The TEI code 1111111 (=127) indicates a broadcast call to all terminals associated with a service-access point. The remaining values (0–126) are used to identify terminals. The range is split between those terminals that use a TEI selected by the network (automatic TEI assignment) and those terminals that use a TEI chosen by the customer (non-automatic TEI assignment).

The control field indicates the type of frame being transmitted, reflecting the acknowledged and unacknowledged types of operation described in Section 4.8.3. There are three types of format defined by the control field: information transfer (I-format), supervisory functions (S-format) and unnumbered transfer (U-format).

The I-format is used to transfer information in the acknowledged type of operation. The I-format control field contains a 'send-sequence number', which is incremented by one (modulo 128) each time a frame is transmitted. There is also a 'receive-sequence number', to acknowledge receipt of previously-received I-frames. The procedure for using the sequence numbers is explained in Section 8.5.3.

The S-format control field is used in the operation of the data-link layer. For example, if a local exchange is temporarily unable to receive any more I-frames, then an S-frame termed 'receive-not-ready' is sent to the customer. When the local exchange can receive I-frames again, a 'receive-ready' S-frame is used. The S-frame can also be used to acknowledge an I-frame. The S-frame contains a receive-sequence number but not a send-sequence number.

The U-format is used to transfer information in the unacknowledged type of operation and to transfer some administrative instructions. The U-format control field does not contain sequence numbers.

A summary of the important commands and responses is given in Table 8.1. The use of the frames is explained in Section 8.5.

The information field contains the information generated by one Layer 3 (e.g. the customer) to be sent to another Layer 3 (e.g. the local exchange). The information field can be omitted if the frame is not relevant to a specific call (e.g. in supervisory frames). If the frame is pertinent to the operation of the data-link layer and Layer 3 is not involved in the formation of the frame, then the relevant information is included in the control field.

The check-bits field, or more formally the 'frame-check-sequence' field, is generated by the data-link layer transmitting the frame. The bits are generated by applying a complex-polynomial algorithm to the rest of the frame. The algorithm is the same as that described in Section 5.4.7. The reverse process is applied by the data-link layer receiving the frame. If the results of the reverse process align with the check bits, then the frame is deemed to have been transmitted without errors. If the results do not align, then an error has occurred during the transmission of the frame.

8.5 Layer 2 procedures

8.5.1 Unacknowledged-information transfer

The procedures for the unacknowledged transfer of information are illustrated in Fig. 8.4. Consider the case when the Layer-3 functions at the local exchange need to transfer information to the Layer-3 functions at the customer's premises. The local-exchange Layer 3 requests the transfer of information by passing a unitdata-request primitive to Layer 2. Layer 2 formulates an unnumbered-information (UI) command frame, enclosing the information to be transferred within the information field. The frame is transmitted, via Layer 1, to Layer 2 at the customer. If it is intended to broadcast the frame to all terminals, the TEI in the address field is coded 127. If a specific terminal is being addressed, i.e. point-to-point working is required, then the TEI is given the code (within the range of 0–126)

Table 8.1 Examples of commands and responses

Format	Commands	Responses	Description
Information transfer (I)	Information	–	Used in the acknowledged type of operation to transfer sequentially numbered frames containing information fields provided by Layer 3
Supervisory (S)	Receive ready	Receive ready	Used to indicate that a data-link layer is ready to receive an I-frame or acknowledge previously received I-frames
	Reject	Reject	Used to request retransmission of the I-frame
Unnumbered (U)	Unnumbered information (UI)		Used in the unacknowledged type of operation to transfer information fields provided by Layer 3
	Set-asynchronous balanced-mode extended (SABME)		Used to set-up an acknowledged type of operation
	Disconnect		Used to terminate an acknowledged type of operation
		Unnumbered acknowledgement	Used to acknowledge and report acceptance of mode-setting commands eg. SABME, disconnect

appropriate to that terminal. Upon receipt of the UI-command frame at the customer Layer 2, the information contained within the information field is delivered to Layer 3 using a unitdata-indication primitive.

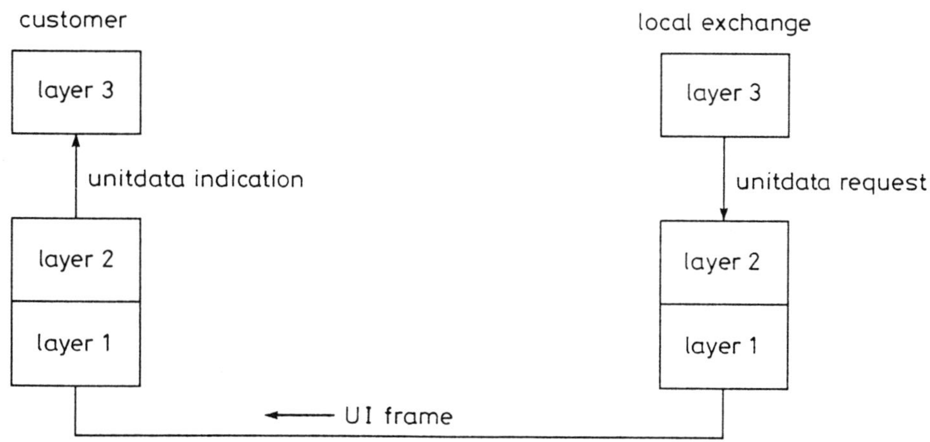

Fig. 8.4   Procedures for unacknowledged – information transfer

In unacknowledged-information transfer, there is no data-link layer error-recovery procedure. Hence, it is left to the Layer 3 functions to determine the logical recovery from the loss or corruption of a frame.

8.5.2 Terminal-endpoint-identifier procedures

Procedures are specified for the assignment, checking and removal of TEIs. All messages used for TEI-management procedures are carried in the information field of UI-command frames.

The assignment procedure allows a customer's automatic-TEI equipment to request a local exchange to assign a TEI value that can be used in subsequent communications. Fig. 8.5 illustrates the principles involved. The customer Layer 2 formulates a command UI-frame. The information field of the UI-frame contains a message consisting of a message-type field, a reference number and an action indicator. The message-type field in this case is called 'identity request', reflecting the request for the local exchange to supply a TEI. The reference number is randomly generated for each request for a TEI. The number is used to discriminate between simultaneous requests for a TEI assignment. The action indicator is used to request the assignment of any appropriate TEI value. The frame is transmitted to Layer 2 at the local exchange, where it is analysed and a suitable TEI value is selected. Layer 2 at the local exchange formulates a command UI-frame. The information field of the UI-frame contains a message consisting of a message type (in this case called 'identity assigned'), the reference number provided by the customer and the assigned TEI value in the action-indicator field. The reference number is used to associate the response with the corresponding request. Several procedures are defined in case the assignment of a TEI value fails [2], but these are not described here.

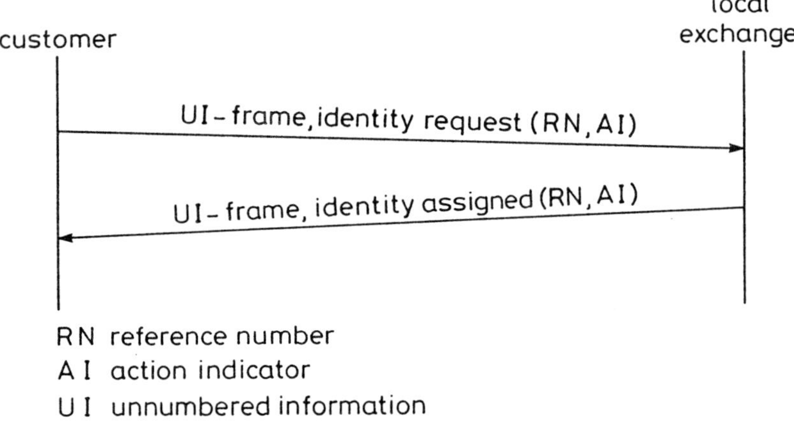

Fig. 8.5   Terminal-endpoint – identifier procedure

The check procedure allows the local exchange to check, for example, whether or not a TEI value is in use. The procedure is illustrated in Fig. 8.6. Layer 2 at the local exchange formulates a command UI-frame consisting of a message type (named 'identity-check request') and an action indicator including the TEI value to be checked. The frame is transmitted to the customer Layer 2.

If any customer equipment recognises the TEI value as its own, it responds by transmitting a UI-frame consisting of a message type (named 'identity-check response') and an action indicator field including the TEI value. A reference number is also included in the response for other, more detailed, purposes. Again, procedures are specified to overcome failures in the logical sequence described. The check procedure can, optionally, be initiated by the customer generating an 'identity-verify' message type. Upon receipt of such a message type at the local exchange, the procedures in Fig. 8.6 are implemented.

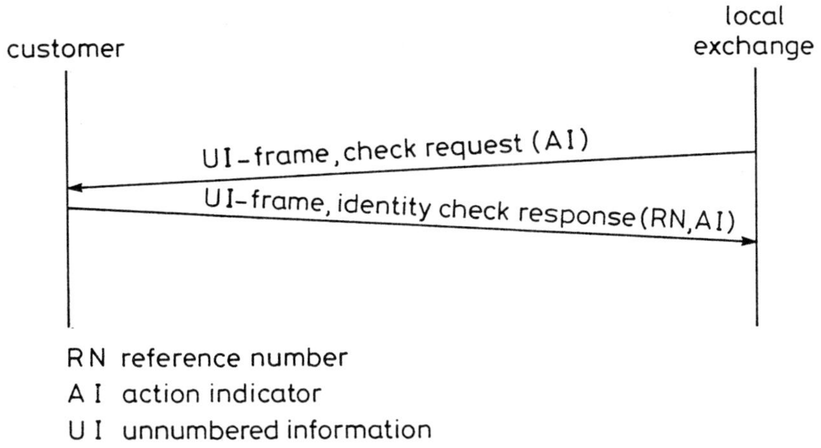

Fig. 8.6   Terminal-endpoint – identifier check procedure

If the local exchange determines that a TEI value should be deleted, the 'removal' procedure is invoked. The local-exchange Layer 2 formulates a frame consisting of a message type (named 'identity remove') and an action indicator field including the TEI value to be removed. The frame is sent twice to reduce the risk of loss. Actions at the customer Layer 2 depend upon local circumstances; however, a typical action is to remove the TEI value and initiate an assignment procedure to request an alternative TEI value.

8.5.3 Acknowledged-information transfer

The procedures for acknowledged-information transfer are illustrated in Fig. 8.7. Consider the case of Layer 3 at the customer needing to initiate the transfer of information with the Layer 3 at the local exchange. The procedures are initiated by the customer Layer 3 sending an 'establish-request' primitive to customer Layer 2. The customer Layer 2 formulates a 'set-asynchronous-balanced-mode-extended (SABME)' command frame, indicating the need to establish a Layer 2 acknowledged- information transfer with the local exchange. The SABME-command frame is transmitted to the local exchange via Layer 1. Upon receipt of the SABME command at the local-exchange Layer 2, the frame is analysed and the conditions required to establish an acknowledged-information transfer are checked (e.g. to ensure that all appropriate equipment is available). Provided that the appropriate conditions are met, the local-exchange Layer 2 sends an 'establish-indication' primitive to Layer 3 to indicate that an acknowledged-information transfer is being established. The local exchange also returns an unnumbered-acknowledgement response to the customer Layer 2. Upon receipt of the unnumbered-acknowledgement response at the customer Layer 2, an 'establish-confirm' primitive is passed to the customer Layer 3 to indicate that acknowledged-information transfer can begin.

Information transfer can now occur between the customer and the local exchange using I-frames. The information to be transferred is supplied by Layer 3 in the form of a 'data-request' primitive. The data is incorporated in the information field of the I-frame and transmitted from the customer to the local-exchange via Layer 1. Upon receipt of the I-frame at the local exchange Layer 2, the data are extracted from the information field and passed to Layer 3 in a 'data-indication' primitive. Depending upon the contents of the received I-frame, the local exchange responds to the customer with either an I-frame, or a receive-ready-supervisory frame. Both frames include an acknowledgement that the I-frame from the customer was received successfully.

Each I-frame and supervisory frame contains a send-sequence number (SSN) and a receive-sequence number (RSN) in the control field to ensure that frames are not lost. The loss-detection procedure works in both directions; however, for explanation purposes, consider the customer sending I-frames to the local exchange. The customer equipment increments the SSN by value 1 (modulo 128) each time an I-frame is sent. Upon receipt of an I-frame at the local exchange, the SSN in the control field is compared with the expected value (i.e. the last received SSN plus 1). If the SSN is the same as the expected value, then the local exchange deduces that no I-frames have been lost. If, however, there is a discrepancy between the SSN and the expected value, then the local exchange recognises that an I-frame(s) has been lost and the received I-frame is discarded.

Similarly, each I-frame received by the customer contains a receive-sequence number (RSN) in the control frame. The RSN is used by the local exchange to acknowledge the receipt of I-frames and is set to the value of the next-expected SSN. In this way, the customer receives an acknowledgement that the I-frames which have been sent have been successfully received by the local exchange. If the local exchange discards a frame (e.g. because of lack of correlation of SSN and expected value), the RSN is not incremented and hence the frame is not acknowledged. After a short time, the customer equipment recognises that a particular frame has not been acknowledged and the frame is re-transmitted.

When the exchange of I-frames is completed, the release procedures that apply are shown in Fig. 8.7. In the Figure, the customer Layer 3 sends a 'release-request' primitive to Layer 2. The customer Layer 2 formulates a disconnect frame which is transmitted, via Layer 1, to the local-exchange Layer 2. Upon receipt of the disconnect frame at the local-exchange Layer 2, a 'release-indication' primitive is passed to Layer 3 and an unnumbered-

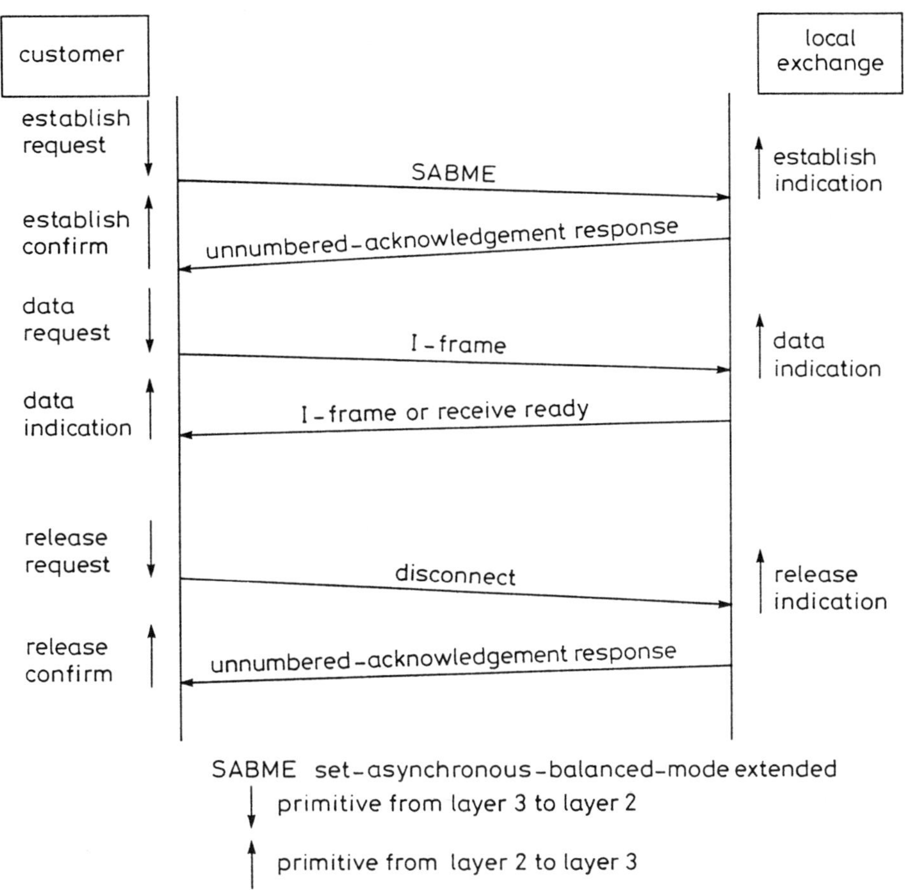

Fig. 8.7   Procedures for acknowledged – information transfer

acknowledgement frame is returned to the customer. Upon receipt of the unnumbered-acknowledgement frame by the customer Layer 2, a 'release-confirm' primitive is passed to Layer 3 to complete the release procedure.

8.6 Chapter summary

It is essential to provide a modern access-signalling capability to take full advantage of the provision of ISDNs. DSS1 has been defined to provide such a capability. DSS1 aligns with the OSI 7-layer model.

Layer 1 of DSS1 defines the physical, electrical and functional characteristics of the transmission link between the customer and the local exchange. Two forms of access are defined. The basic access is formed of two B channels (each of 64 kbit/s) and a D channel of 16 kbit/s. The D channel is the signalling channel. The primary access is formed of thirty B channels and a D channel of 64 kbit/s.

Layer 2 of DSS1 is responsible for transferring information between the customer and the local exchange. Information is transferred in frames, which are equivalent to the signal units of CCITT No. 7. Frames consist of flags, an address field, a control field, an information field and check bits. The flags delimit the frame. The address field identifies the receiver of a command frame or the sender of a response frame. The control field indicates the type of frame being transmitted (I-format, S-format or U-format). The information field contains the information supplied by Layer 3 to be transferred. The check bits are used to detect errors during transmission of the frame.

Two forms of transfer are defined. In unacknowledged-information transfer, information is exchanged in unnumbered-information frames without sequencing. In this case, there is no error-correction method defined at Layer 2. In acknowledged-information transfer, a Layer 2 connection is established by using a set-asynchronous-balanced-mode-extended frame and information is exchanged using sequenced-information frames. The connection is cleared by using a disconnect frame.

8.7 References

1 CCITT Recommendations Q.920, 921, 930 and 931: 'Digital Subscriber Signalling System No. 1' (ITU, Geneva)
2 CCITT Recommendation Q.921: 'ISDN user-network interface, data-link layer specification' (ITU, Geneva)
3 ISO Standard 4335: 'Data communication, high-level data-link control procedures'

Chapter 9

DSS1 network layer

9.1 Introduction

The DSS1 network layer (Layer 3) is responsible for establishing, maintaining and clearing circuit-switched traffic channels between customers and ISDN local exchanges. It is also responsible for providing access to packet-data facilities. It is Layer 3 of DSS1 that formulates and analyses the messages carried by Layers 1 and 2. In this role, the DSS1 network layer interacts with the customer and translates the customer requirements into messages to be sent to an ISDN local exchange. At the local exchange, the DSS1 network layer interacts with the call-control function to stimulate actions meeting the customer requirements. The specifications of DSS1 Layer 3 [1,2] adopt the term 'user' to describe a customer. This chapter continues with the term customer unless reference is made to a specified message or information (e.g. user-to-user information).

DSS1 Layer 3 can be described in terms of its formats and the procedures, defining the logical sequence of events, in meeting customer requirements. Section 9.2 describes the principles of the format technique and Section 9.3 gives examples of messages to illustrate the principles. The basic procedures for establishing circuit-switched calls are described in Section 9.4 and the procedures for clearing such calls are described in Section 9.5. DSS1 Layer 3 includes features additional to those necessary for establishing and clearing calls and these features are outlined in Section 9.6. DSS1 Layer 3 is required to provide access to packet-data facilities and the means of achieving this requirement are described in Section 9.7. Section 9.8 covers the invocation and transfer of user-to-user signalling over the access signalling channel. Finally, whilst the DSS1 network layer does not specify detailed procedures for the control of supplementary services, three generic procedures are outlined in Section 9.9.

9.2 Format principles

The general format for DSS1 Layer 3 messages is shown in Fig. 9.1. Each message contains a protocol discriminator, a call reference and a message type. Other information elements are defined according to the type of message.

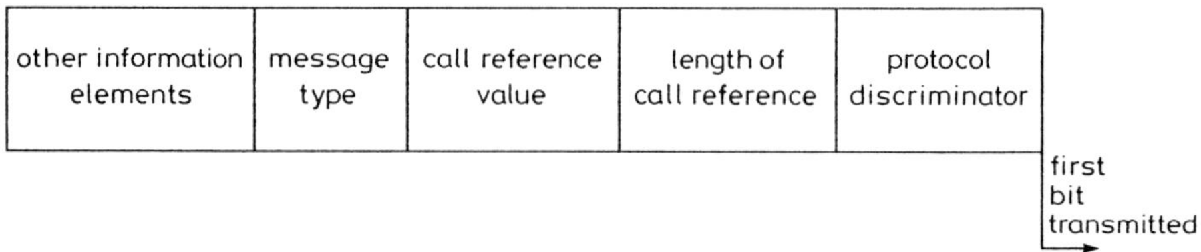

Fig. 9.1   General format of DSS1 Layer 3 messages

9.2.1 Protocol discriminator

The first part of every message is the protocol discriminator, consisting of one octet. The purpose of this field is to distinguish between DSS1 messages for call control and any other type of message that can be transported over the signalling channel. For example, it is possible to transmit packet data (explained in Chapter 4) over an access-signalling channel and it is essential that a local exchange can distinguish between DSS1 messages and packet data. The protocol discriminator is coded to ensure that a distinction can be made. The protocol discriminator is coded as 00001000 for circuit-switched call-control messages.

9.2.2 Call reference

The call reference is a number that is used to identify the call to which a message refers. The call reference is assigned during set-up from a pool of numbers and remains fixed for the duration of the call. At the end of the call, the call reference is returned to the pool of numbers in readiness to be used again. The use of a call reference to identify a particular transaction signifies that DSS1 can be non-circuit-related. Circuit-identification information is included in appropriate messages to allow correlation of messages with the corresponding traffic channel.

The format of the call-reference-information element is illustrated in Fig. 9.2. The first four bits of the first octet indicate the length of the call-reference value. The remaining bits of the first octet are spare. For basic access, the length of the call-reference value is at least one octet and for primary-rate access, the length is at least two octets.

The call reference is assigned either by the customer equipment or by the local exchange. If the customer originates a call, then the customer assigns the call reference from its pool of numbers. Similarly, if the local exchange originates a call (in response to a request from the network), then the local exchange selects the call reference. With this approach, it is possible that both the customer and the local exchange select the same call reference value for different calls. To ensure that it is possible to discriminate between the two calls in this case, a 'flag' is included as the last bit of the first octet of the call-reference value. The flag indicates whether the customer or local exchange originated a particular call-reference value. Note that the flag in this context is not the same as the flag used to delimit frames.

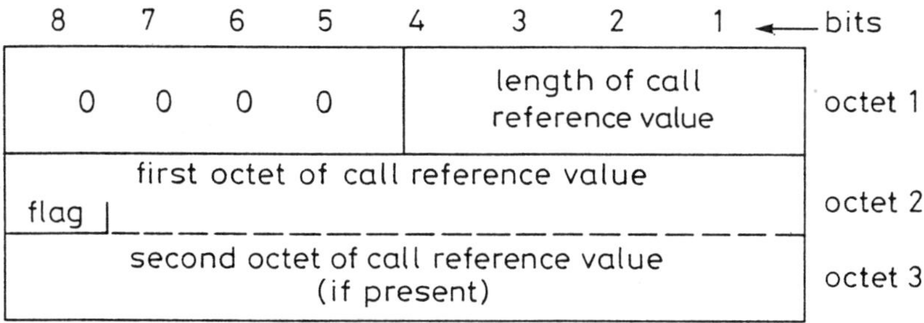

Fig. 9.2   Format of call-reference information element

9.2.3 Message type

The message type is used to identify the function of the message being sent. The message-type field consists of one octet, with the last bit being reserved for use in the future to extend the length of the field. There are five categories of message type:

(*a*)   Call-establishment messages, that deal with the set-up of calls. For example, the 'set-up' message is sent by a calling customer to the local exchange to initiate a call.

(*b*)   Call-information-phase messages, that are used when a call has been established. For example, a 'user-information' message can be sent during the conversation/data phase of a call to transfer user-to-user information.

(*c*)   Call-clearing messages, that deal with the clear-down of calls. For example, the 'disconnect' message is sent by the customer to the local exchange to clear a connection.

(*d*)   Miscellaneous messages, e.g. an 'information' message can be sent by the customer or the local exchange to supply information additional to that provided in other basic messages.

(*e*)   Nationally-specified messages. The message-type code 00000000 is used to denote that the next field is a message type that is defined by a network operator. Thus, the next field in a message of this type is not standardised internationally.

9.2.4 Other-information elements

Two categories of information element are defined, depending on whether they are one octet in length (single-octet-information elements) or more than one octet in length (variable-length-information elements).

There are two types of single-octet-information element. Type 1 is illustrated in Fig. 9.3. Bit 8 (value 1) indicates that the element belongs to the single-octet category and bits 5–7 are used to define the name of the information element in the form of an 'information-element identifier'. Bits 1–4 provide the contents of the information element. Type 2 is illustrated in Fig. 9.4. Again, Bit 8 (value 1) is used to show that the information element belongs to the single-octet category. However, the remainder of the octet is used entirely as the information-element identifier.

8	7	6	5	4	3	2	1 ←bits
1	information–element identifier			contents of information–element			

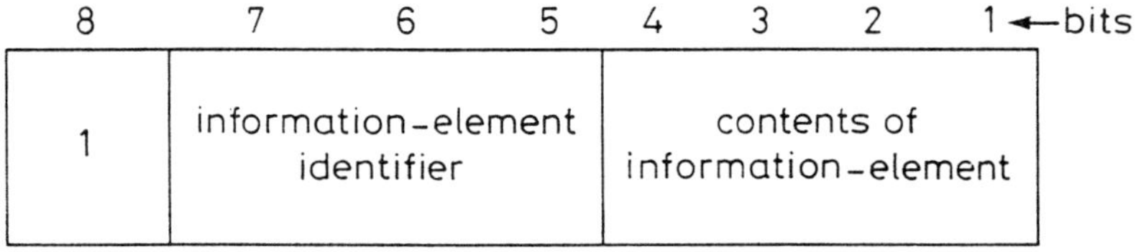

Fig. 9.3   Type 1 single-octet-information element
Source: CCITT Recommendation Q.931

8	7	6	5	4	3	2	1 ←bits
1	information–element identifier						

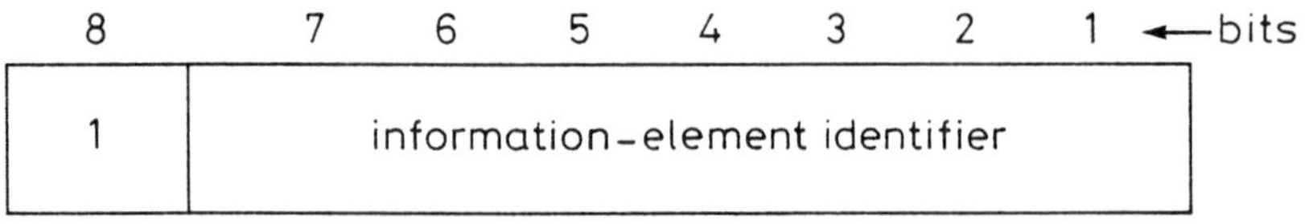

Fig. 9.4   Type 2 single-octet-information element
Source: CCITT Recommendation Q.931

The format of the variable-length-information element is shown in Fig. 9.5. In this case, Bit 8 of the first octet is set to value 0, distinguishing this category of element from the single-octet category. The remainder of the first octet is dedicated to the information-element identifier. The second octet defines the length of the contents of the information element and the third and subsequent octets give the contents.

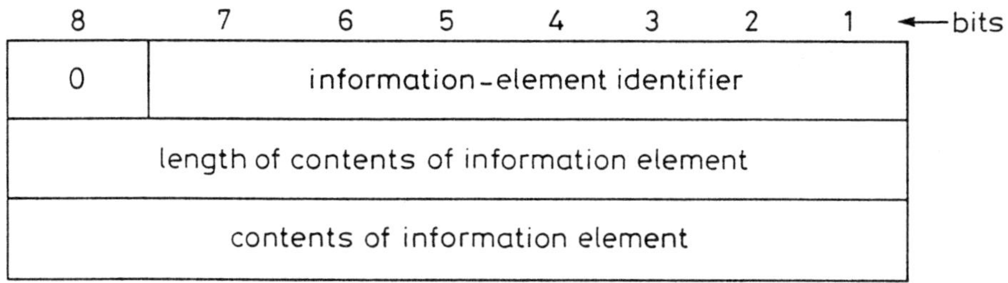

Fig. 9.5   Variable-length-information element
Source: CCITT Recommendation Q. 931

Single-octet-information elements can be placed at any point within the information-element field of a message. However, variable-length information elements are placed in ascending order of the information-element identifier. This allows equipment at the customer's premises, or at the local exchange, to detect the presence or absence of a particular piece of information without having to analyse the entire message.

9.2.5 Codesets

The format of the variable-length-information element allows 7 bits for the information-element identifier. Thus, up to 128 different information elements can be identified in this category. The number of bits allowed in single-octet-information elements depends upon the type adopted, with 3 bits being available in Type 1 and 7 bits in Type 2. Thus, at least 8 different information elements can be identified in this category. Combining the two categories, at least 136 information-element identifiers are available, although in practice this is reduced to 133 after allowing for reserved values. This group of 133 information-element identifiers is termed a 'codeset'.

The number of information elements that can be identified within a message can be increased substantially by re-using the information element-identifier codes in different codesets. Thus, a given information-element-identifier code can define one information element in one codeset and another information element in another codeset. The principle is illustrated in Fig. 9.6. Consider the information-element identifier in Codeset 0, termed 'calling-party number' and coded 1101100. The same code, 1101100, can be used in another codeset (Codeset 5 in Fig. 9.6) to define a completely-different information element. Codeset 5 is reserved for network operators to use for non-standard codings of information elements: in this example, the code 1101100 is used to define the 'time' information element, giving the time at which a message is sent. The same code can be used again in other codesets to define other information-element identifiers.

codeset 0	
information–element identifier	
code	name
•	•
•	•
•	•
1101100	calling–party number
•	•
•	•
•	•

codeset 5	
information–element identifier	
code	name
•	•
•	•
•	•
1101100	time
•	•
•	•
•	•

Fig. 9.6   Re-use of information-element identifier codes

For this re-use approach to work, it is essential that any equipment sending or receiving a message is fully aware of the codeset being employed. This applies to equipment at the customer's premises and to equipment at the local exchange. The method adopted by DSS1 [1] is to define Codeset 0 as the initial codeset used for analysing a message. If any equipment needs to use an information element within a different codeset (e.g. Codeset 5) then the frame of reference is 'shifted' from Codeset 0 to Codeset 5. This is achieved by transmitting a Type 1 single-octet-information element termed 'shift', which is illustrated in Fig. 9.7. Bit 8 of the shift-information element is coded 1 to indicate a single-octet-information element. Bits 5 to 7 are the information-element identifier and are coded 001 to define the request to shift from the current codeset to a new codeset. The new codeset is defined in Bits 1-3.

<table>
<tr><td>8</td><td>7   6   5</td><td>4</td><td>3   2   1</td><td>←bits</td></tr>
<tr><td>1</td><td>shift identifier</td><td>lock/
temporary
bit</td><td>new codeset
identification</td><td></td></tr>
</table>

Fig. 9.7   Shift-information element

Bit 4 is used to indicate whether the shift in codeset should be applied for the remainder of the message (locking-shift procedure) or temporarily (non-locking-shift procedure). In the locking-shift procedure, the new codeset in the shift-information element is applied for the remainder of the message (or until a further shift is requested). For example, Codeset 0 is used as the standard codeset in the analysis of a message. If a locking-shift-information element is received requesting a shift to Codeset 5, the remaining information elements of the message are interpreted according to the information-element identifiers in Codeset 5 (unless further shift information is received). In the non-locking-shift procedure, the new codeset in the shift-information element is only applied to the next information element. Thus, if a non-locking-shift information element is received requesting a shift to Codeset 5, only the next information element is interpreted according to the information-element identifiers in Codeset 5. After this information element is interpreted, the codeset used to interpret subsequent information elements reverts to Codeset 0.

The shift-information element illustrated in Fig. 9.7 allocates 3 bits for the new codeset identification. Thus, up to 8 codesets can be accommodated. CCITT has specified [1] three values of codeset in addition to Codeset O. Codeset 5 is reserved for national use, thus allowing network operators to use codings that do not form part of the international specification. Codeset 6 is reserved for local networks, e.g. between customers and local exchanges. Codeset 7 is reserved for customer-specific information, i.e. the information is applicable only to customers and is not recognised by local exchanges or other parts of the network.

9.3 Examples of message formats

9.3.1 General

The principles of the format technique for DSS1 are illustrated below by selecting some commonly-used message types and information elements. The messages can be relevant to different portions of a call. 'Global significance' is used to describe messages that are relevant to all portions of a call, i.e. the link between the originating customer and local exchange (originating access), the link between the terminating customer and local exchange (terminating access) and within the network. 'Dual significance' is used to describe messages relevant to the originating access and the network or the terminating access and the network. 'Access significance ' means that messages are relevant only to the originating and terminating accesses. Finally, 'local significance' is used to describe messages relevant to either the originating access or the terminating access.

The information elements in the examples are all from Codeset 0. Each information element is either mandatory (i.e. it must be included in a particular message) or optional (i.e. it can be included or not, as appropriate). The length in octets of each information element is also given. The message-type examples chosen are appropriate for the control of both circuit-switched and packet-data calls. Other messages can be applicable to only one type of call control. Similarly, the full range of information elements included in one type of call control is not necessarily applicable to the other type of call control.

9.3.2 Set-up message

An example of the format of the set-up message is given in Table 9.1. This message is sent from the calling customer to the originating local exchange and from the terminating local exchange to the called customer. The message initiates call establishment and is equivalent to the initial-address message in the ISUP of CCITT No. 7 (Chapter 6). In circuit-switched call control, the set-up message has global significance and includes compatibility information elements that are used to ensure that the calling and called customer terminals can communicate effectively. For example, a calling customer wishing to make a telephone call should not be connected to a data terminal at the called customer.

Table 9.1 Example of set-up message

Information element	Type	Length	Description
Protocol discriminator	M	1	See Section 9.2.1
Call reference	M	$\geqslant 2$	See Section 9.2.2
Message type	M	1	See Section 9.2.3
Bearer capability	M	4–13	Defines a large range of features relating to the type of connection and service required by the customer. Examples are whether the call is circuit-switched mode or packet-data mode, the bandwidth required and the information-transfer rate
Channel identification	O	$\geqslant 2$	Identifies the channel to which the message pertains. Mandatory in network-to-customer direction
Progress indicator	O	2–4	Used to describe an event that occurs during the call e.g. to inform the destination customer that the call is not an end-to-end ISDN call
Display	O	2–82	Provides information that can be displayed by the customer terminals
Calling-party number	O	$\geqslant 2$	Gives the address of the calling customer
Called-party number	O	$\geqslant 2$	Gives the address of the called customer
User-user	O	2–131	Used to convey user-to-user information

M : mandatory
O : optional

9.3.3 Connect message

The connect message denotes that a call is accepted by the called customer. It is sent by the called customer to the network and by the network to the calling customer. The message has global significance for circuit-switched call control and local significance for packet-data call control. The message is the equivalent of the answer message in the ISUP of CCITT No. 7. An example of the format of the message is given in Table 9.2.

Table 9.2 Example of connect message

Information element	Type	Length	Description
Protocol discriminator	M	1	See Section 9.2.1
Call reference	M	$\geqslant 2$	See Section 9.2.2
Message type	M	1	See Section 9.2.3.
Channel identification	O	$\geqslant 2$	Identifies the channel to which the message pertains. Can be mandatory if connect is the first response to a set-up message
Progress indicator	O	2–4	As for set-up message
Display	O	2–82	As for set-up message
User-user	O	2–131	Used to convey user-to-user information, but only for circuit-switched call control

M : Mandatory
O : Optional

9.3.4 Disconnect message

The disconnect message is sent by a customer (either the calling or called customer) to request the network to clear the call. It is also sent by the network to the other customer to indicate that the call is cleared. The message has global significance for circuit-switched call control and local significance for packet-data call control. An example of the format of the message is given in Table 9.3.

9.4 Basic call-establishment procedures for circuit-switched calls

9.4.1 General

The procedures for the establishment of basic circuit-switched calls assume that a Layer-2 data-link connection (Chapter 8) is in operation between the calling customer and the originating local exchange before a call is initiated. Such a data-link connection might not be in operation between the destination local exchange and the called customer when a request to set-up a call is received. However, it is assumed that a data-link connection is established before any response is given to a request to set-up a call.

Table 9.3 Example of disconnect message

Information element	Type	Length	Descripton
Protocol discriminator	M	1	See Section 9.2.1
Call reference	M	$\geqslant 2$	See Section 9.2.2
Message type	M	1	See Section 9.2.3
Cause	M	4–32	Gives the reason for disconnecting (eg. normal customer clearing or destination out of order) and the location of the initiator of the message (eg. customer or network).
Display	O	2–82	As for set-up message
User-user	O	2–131	Used to convey user-to-user information for circuit-switched call control and in some circumstances for packet-data call control

M : Mandatory
O : Optional

The procedures vary according to whether en-bloc or overlap operation (explained in Chapter 6) is adopted by the calling customer. They also vary according to whether the called customer has multiple terminals associated with a signalling connection or a single terminal. If the customer has multiple terminals, the broadcast form of working of the data-link layer (explained in Chapter 4) is used to initiate a call. If the destination local exchange is aware that a single terminal exists, and the terminal endpoint identifier is known, the point-to-point form of working is used.

9.4.2 En-bloc procedures in conjunction with point-to-point working

In en-bloc operation, a call is initiated by the calling customer formulating a set-up message and sending it to the originating local exchange over the signalling channel (D channel). The procedures are illustrated in Fig. 9.8. The set-up message includes all the information needed to route the call to the called customer and a call reference, chosen from a pool of numbers, is used to correlate all messages between the calling customer and the originating local exchange for that call. The set-up message also includes a bearer-capability-information element (defining the type of connection required to fulfil the needs of the calling customer) and information to check that the calling and called customer terminals are compatible. The customer indicates, in the channel-identification-information element, either the identity of the traffic channel upon which the call should be made or an indication that any traffic channel will suffice. If a specific traffic channel is nominated, it can be qualified as being essential or desirable.

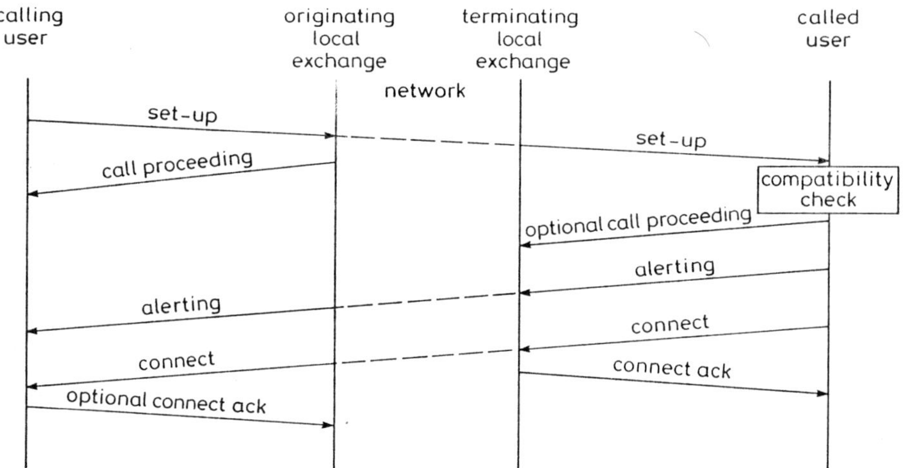

Fig. 9.8   Basic call-establishment procedures: circuit-switched, en-bloc

Upon receipt of the set-up message from the calling customer, the originating local exchange returns a 'call-proceeding' message to acknowledge receipt of the set-up message and indicate that the call is being processed. The call-proceeding message confirms the traffic channel to be used for the call. The originating local exchange also transmits the request to establish a call through the network to the terminating exchange.

Upon receipt of a request to set-up a call, the terminating local exchange sends a set-up message to the called customer over the D channel. In this example, the point-to-point data-link layer form of working is used. The set-up message includes a call reference generated by the terminating local exchange and a proposed traffic channel that should be used for the call. The set-up message also includes the compatibility-check information supplied by the calling customer. Before any further action is taken, the compatibility information is checked by the called customer to ensure that effective communication can take place between the calling and called customers. If compatibility is not achieved, the call is rejected by the called customer returning a 'release-complete' message with an information element giving the reason for rejecting the call (e.g. incompatible terminals). If compatibility is achieved, the call set-up process continues.

At this stage, the called-customer equipment can return a call-proceeding message to the terminating local exchange, but this message is optional. The message informs the terminating local exchange that the call is being processed by the customer equipment. However, the message results in no specific action at the local exchange other than to reset timers that are used by the local exchange to check for faulty conditions (e.g. non-response to a set-up message).

The next step in the set-up sequence depends upon the type of terminal connected at the called customer's premises. Some terminals automatically answer an incoming call without manual intervention (e.g. some data terminals) and these are termed 'automatic-answering' terminals. Other terminals need manual intervention to answer a call (e.g. the lifting of the handset on a telephone) and these are termed 'non-automatic-answering' terminals.

For a non-automatic-answering terminal, an 'alerting' message is returned to the terminating local exchange, indicating that the called customer is being notified that there is an incoming call. This is equivalent to ringing the telephone in a telephony call, thus notifying the called customer that an incoming call is being offered. The terminating local exchange passes the alerting information through the network to the originating exchange. The originating exchange sends an alerting message to the calling customer to keep the calling customer informed of call progress. When the call is answered by the called customer (e.g. the telephone handset is lifted), a 'connect' message is sent from the called-customer terminal to the terminating local exchange. This message is passed through the network to the originating local exchange and subsequently to the calling customer. The terminating local exchange sends a connect-acknowledge message to the called customer to confirm that the call is now established. Optionally, the calling-customer terminal can send a connect-acknowledge message to the originating exchange.

For an automatic-answering terminal, the speed of response to an incoming call is generally much faster than for a non-automatic-answering terminal. Whereas for a non-automatic answering terminal there can be expected to be an extended alerting period, this does not occur in general with an automatic-answering terminal. Thus, for an automatic-answering terminal, the alerting message can be omitted from the set-up sequence. This means that the response to set-up message from an automatic-answering terminal is a connect message (if the optional call-proceeding message is not used).

The set-up message to the called customer includes the identity of the traffic channel which it is proposed to use for the call. If possible, the customer terminal selects the identified traffic channel. However, if it is not possible to select that channel, the customer terminal chooses another traffic channel and informs the terminating local exchange in the first response to the set-up message (i.e. the call-proceeding, alerting or connect message).

9.4.3 En-bloc procedures in conjunction with broadcast working

The procedures adopted with broadcast working are similar to those described for point-to-point working in Section 9.4.2. The differences, described below, are to cater for responses from multiple terminals. In this case, the set-up message from the terminating local exchange to the called customer is sent in broadcast form, i.e. it is addressed to all terminals associated with the signalling channel (D channel). Each terminal checks the compatibility information supplied in the set-up message to determine whether or not it is compatible with the calling-customer terminal. If a terminal determines that it is not compatible, then it can either ignore the set-up message and take no further action or it can return a release-complete message with a 'cause' information element indicating that it is not compatible with the calling-customer terminal. If a terminal determines that it is compatible with the calling-customer terminal, then it responds to the terminating local exchange with a call-proceeding message, an alerting message and/or a connect message, according to the criteria described in Section 9.4.2. The terminating local exchange needs to keep track of each terminal that responds.

The first called-customer terminal to respond to the terminating local exchange with a connect mesage is deemed to be the recipient of the call. A connect-acknowledge message is sent from the terminating local exchange to the recipient terminal to confirm that it has been selected to take the call. The recipient terminal can then connect to the appropriate traffic channel. The terminating local exchange also sends a release message to every terminal that responded to the set-up message (expcet the recipient teriminal) to indicate that the offer of the call has been withdrawn.

In the broadcast form of working, negotiation of the traffic channel to be used between the terminating local exchange and the called customer is not possible, unless specialised procedures are adopted. Thus, the call is made using the traffic channel selected by the terminating local exchange.

9.4.4 Overlap procedures

Overlap operation can be adopted in the originating access, the terminating access, within the network and in combinations of these three elements. To illustrate the principles involved, consider the case when the originating access and the network use overlap operation and the terminating access uses en-bloc operation. The procedures in this case are illustrated in Fig. 9.9. The contents of the messages are similar to those described in Section 9.4.2 and they are not repeated here. Differences in procedures between point-to-point working and broadcast working are similar to those described in Section 9.4.3.

Overlap operation is initiated by the calling customer sending a set-up message to the originating local exchange that does not include the full address of the called customer. The originating local exchange returns a 'set-up-acknowledge' message to the calling customer, indicating that further address information is required. Further address digits are provided to the originating local exchange in one or more 'information' messages. When sufficient address digits have been received to route the call to the terminating local exchange, the request to set-up a call is transmitted to the terminating local exchange in an ISUP initial-address message. In the meantime, further address digits can be received by the originating local exchange from the calling customer in information messages: these are passed through the network to the terminating local exchange.

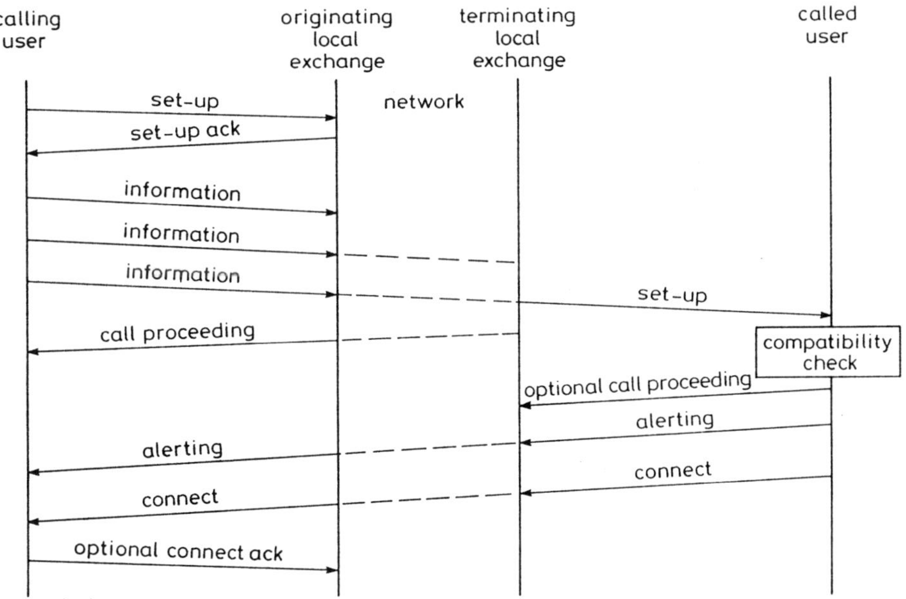

Fig. 9.9 Basic call-establishment procedures: circuit switched, overlap

When the terminating local exchange has received sufficient address information to identify the called customer, a set-up message is sent to initiate the call-establishment procedures on the terminating-access link. The procedures on the terminating-access link are the same as those described in Sections 9.4.2 and 9.4.3. The terminating local exchange also informs the originating local exchange that sufficient address information has been received. In response to this information, the originating local exchange sends a call-proceeding message to the calling customer. The call-proceeding message indicates that no further address information will be accepted. Procedures beyond this stage of the call are covered in Sections 9.4.2 and 9.4.3.

9.5 Call-clearing procedures for circuit-switched calls

In these procedures, the term 'disconnected' means that the appropriate traffic channel has been cleared-down, but it is not yet ready to be used again for another call. The term 'released' means that the traffic channel has been cleared-down and is available to be used again for another call. Similarly, when the term 'released' is applied to a call reference, the call reference is available to be used again for another call.

The procedures for clearing a call are illustrated in Fig. 9.10. The clearing procedures can be initiated by either the calling customer or the called customer: this example assumes that the calling customer requests the call to be cleared. In this case, the calling customer transmits a disconnect message over the signalling channel (D channel) to the originating local exchange. The calling customer also disconnects the traffic channel (B channel). Upon receipt of the disconnect message, the originating local exchange disconnects the traffic channel and, when this is completed, returns a release message to the calling customer. The originating exchange also passes the request to clear the call to the terminating exchange through the network. When the calling customer receives the release message, the traffic channel is released, a release-complete message is sent to the originating exchange and then the call reference is released. Upon receipt of the release-complete message, the originating exchange releases the traffic channel and the call reference. Note that whilst the traffic channel is first disconnected and then released, the call reference is held active until the last message in the clear sequence is sent. This is essential because the call reference is needed to correlate the messages with the call: the final act is to release the call reference.

When the terminating local exchange receives the request to clear the call from the originating exchange, the terminating local exchange disconnects the traffic channel to the called customer and sends a disconnect message over the signalling channel. This message prompts the called customer to disconnect the traffic channel and return a release message to the terminating local exchange. Upon receipt of the release message, the terminating local exchange releases the traffic channel, sends a release-complete message to the called customer and releases the call reference. Receipt of the release-complete message causes the called customer to release the traffic channel and call reference.

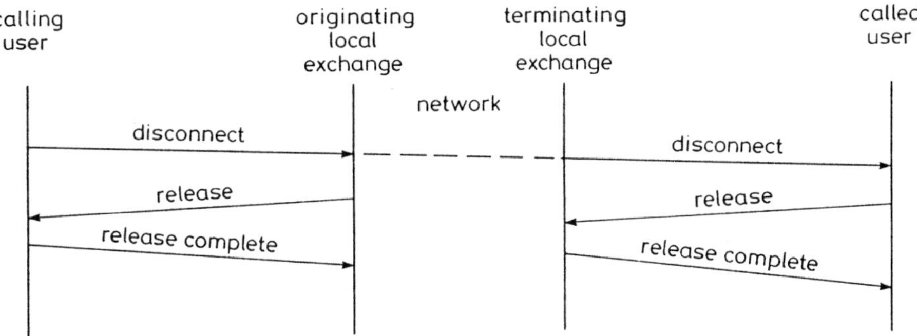

Fig. 9.10   Call-clearing procedures: circuit-switched calls

9.6 Other features for basic circuit-switched calls

9.6.1 General

The specification of the basic procedures for DSS1 [1] includes a range of features additional to call establishment and call clearing. A summary of some of the important features is given below.

9.6.2 Restart procedure

If a fault occurs on an access link, the customer and local exchange can lose track of the status of the channels in that access link. The restart procedure can be used to return the channels to an idle condition. The procedure can also be invoked if, for example, a customer terminal fails to respond to the clearing sequence.

The procedure is initiated by either the customer or the local exchange sending a 'restart' message. If the restart applies to one channel, then the message includes a channel-identification-information element. Upon receipt of a restart message, the recipient clears the relevant channel(s) and call reference(s) and returns a 'restart-acknowledge' message. Receipt of the restart-acknowledge message causes the initiator of the procedure to release the appropriate channel(s) and call reference(s).

9.6.3 Call-rearrangement procedure

This procedure allows a customer to suspend a call, make changes to the terminal being used and then resume the call. The changes can include physically replacing one terminal with another, physically moving from one terminal to another and disconnecting and reconnecting a terminal.

The procedure is initiated by the customer sending a 'suspend' message to the local exchange. The message includes a call identity that replaces the call reference, thus allowing the local exchange to release the call reference. The local exchange confirms suspension by returning a 'suspend-acknowledge' message. When the customer wishes to resume the call, a 'resume' message is sent to the local exchange, including a new call reference and the call identity. The local exchange correlates the call identity with that included in the suspend message and returns a 'resume-acknowledge' message to allow the call to continue.

9.6.4 Error conditions

A range of procedures to handle error conditions is specified. This is essential to ensure that the customer and local exchange have procedures to cover any error, or corrupted message, that might occur. For example, if a customer receives an unexpected message, a 'status' message can be returned with an information element indicating that the received message was not understood.

9.7 Procedures for packet-data calls

9.7.1 General

A description of packet-data calls is given in Chapter 4. Whilst details of packet-data calls are beyond the scope of this book, it is worthwhile giving a brief outline of the role of DSS1 in supporting packet communications in an ISDN. There are two ways of providing packet communications. First of all, some network operators have implemented packet-switched public data networks (PSPDNs). These are networks dedicated to the provision of packet communications, just as telephony networks were primarily dedicated to the provision of telephony. PSPDNs use protocols [3,4] optimised for packet data. Secondly, some ISDNs provide an inherent packet-data capability by including equipment, i.e. 'packet handlers', within exchanges. A DSS1 customer can access packet facilities by 'circuit-switched access' or by 'packet-switched access'.

9.7.2 Circuit-switched access

In circuit-switched access, the customer gains access to a PSPDN by a circuit-switched channel. In this case, a circuit-switched traffic channel is established between the customer and a PSPDN. The point at which access is gained to the PSPDN is termed the 'access unit'. The configuration is illustrated in Fig. 9.11.

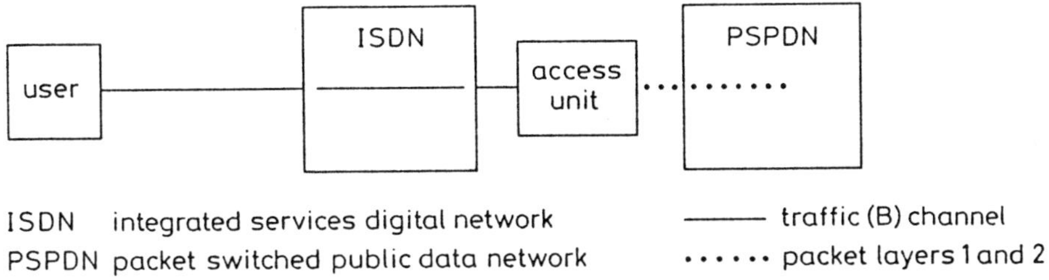

ISDN   integrated services digital network   ——— traffic (B) channel
PSPDN packet switched public data network   •••••• packet layers 1 and 2

Fig. 9.11   Configuration of circuit-switched access to packet network

The procedures adopted on the access link and within the ISDN are the same as for a normal circuit- switched call. Thus, as far as the access-signalling channel (D channel) is concerned, the procedures explained in Section 9.4 are applied to establish a traffic channel between the customer and the access unit. The set-up message includes the called number of the appropriate access unit and the bearer-capability-information element includes an indication that the call is being established in circuit-switched mode. Once the traffic channel has been established, the protocols adopted for the PSPDN are used over the traffic channel between the customer and the PSPDN. The procedures described in Section 9.5 are used to clear the traffic channel upon completion of the call.

9.7.3 Packet-switched access

In packet-switched access, the customer gains access to a packet handler, for example in a local exchange. The customer can decide whether to use a traffic channel (B channel) or the signalling channel (D channel) to gain access to the packet handler.

If access is gained using the traffic channel (B channel), as shown in Fig. 9.12, then the signalling procedures adopted on the D channel are those necessary to establish a B channel between the customer and the packet handler. The procedures are the same as those described in Section 9.4, except that:

(*a*)   En-bloc operation is used and the procedures for overlap operation do not apply

(*b*)   An alerting message is not sent from the local exchange to the calling customer to indicate that the called customer is being alerted

(*c*)   The bearer-capability-information element in the set-up message includes appropriate values, e.g. indicating packet-data mode.

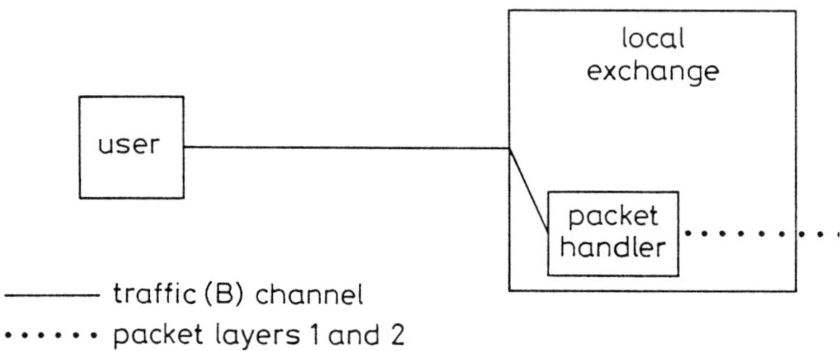

Fig. 9.12   Configuration of packet access using B channel

Once established, the traffic channel is used to convey packet protocols between the customer and the packet handler. The procedures described in Section 9.5 are used to clear the B channel upon completion of the call.

If access is gained using the signalling channel (D channel), as shown in Fig. 9.13, the D channel provides a data-link-layer connection between the customer and the packet handler in the exchange. The data-link layer provides an acknowledged-information-transfer service (I-frames), as explained in Chapter 8. Thus, packet protocols can be conveyed directly between the customer and the packet handler over the D channel. This approach is based on the D channel providing the services of Layers 1 and 2 of the OSI model (see Chapter 4) and the packet protocol providing the Layer-3 service. Similarly, the Layer-3 clearing procedures are those adopted for the packet protocol, whereas the Layer 2 clearing procedures are the standard DSS1 procedures described in Chapter 8.

9.8 User-to-user signalling procedures

9.8.1 General
User-to-user information is exchanged between the calling and called customers without the information being analysed by the network. The network transports the information transparently, as described in Chapter 6 for the ISUP. Three services are offered:

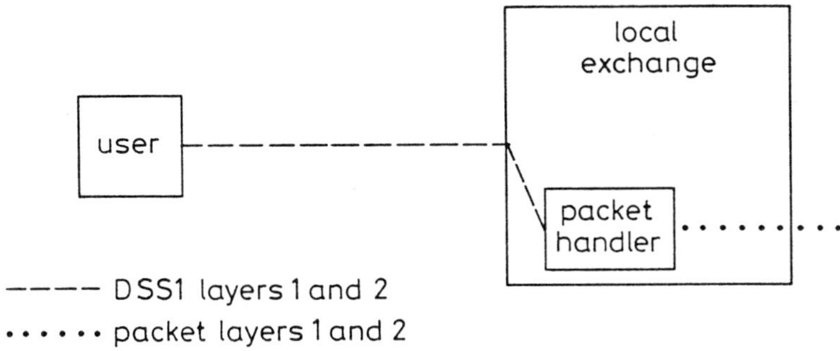

---- DSS1 layers 1 and 2
...... packet layers 1 and 2

Fig. 9.13   Configuration of packet access using D channel

(*a*)   *Service 1:* exchange of user-to-user information during the set-up and clearing phases of a call

(*b*)   *Service 2:* exchange of user-to-user information during call establishment between the alerting and connect messages

(*c*)   *Service 3:* exchange of user-to-user information during the conversation/data phase of the call.

These services are equivalent to those defined in the ISUP. Invocation and operation of the services are described in Sections 9.8.2 to 9.8.4.

9.8.2 Service 1

In Service 1, the user-to-user information is contained in normal call-control messages, e.g. in a set-up message establishing a call. The information is included in a 'user-to-user' information element. Invocation of Service 1 during call establishment can be 'implicit' or 'explicit'.

In the explicit case, the calling-customer specifically states in the set-up message that the message includes user-to-user information. The specific statement is included in a 'facility' information element. The user-to-user information itself is included in a user-to-user information element. The user-to-user information is transferred across the network and delivered to the called customer. Upon receipt of such a set-up message, the called customer responds to the invocation in the alerting or connect messages. The called customer can also include a user-to-user information element in the alerting and/or connect messages. The explicit case (during call establishment) is only applicable for the point-to-point form of working.

In the implicit case, the calling customer includes a user-to-user information element in the set-up message used to initiate the call. However, a specific request to invoke Service 1 is not included (i.e. a facility-information element does not specifically request Service 1). The user-to-user information is transferred across the network and delivered to the called customer. The implicit case applies to both the point-to-point and broadcast forms of working. In the point-to-point form of working, a responding user-to-user information element is included in either the alerting message or the connect message. In broadcast working, the response is included in the connect message.

Service 1 also allows user-to-user information to be included in the first message of the clearing sequence.

9.8.3 Service 2

Service 2 is applicable to the point-to-point form of working. It allows the transfer of user-to-user information between the alerting and connect messages during call establishment. However, there must be an indication in the set-up message that Service 2 is required. The service is invoked by the calling customer including a facility-information element, with an indicator requesting Service 2, in the set-up message. Upon receipt of such a set-up message, the called customer accepts the invocation in the alerting message. Once accepted, user-to-user information can be exchanged between the calling and called customers. The transfer is achieved by 'user-information' messages and two such messages can be exchanged within Service 2. The messages include the relevant protocol discriminators, call references, and user-to-user information elements.

9.8.4 Service 3

Service 3 can either be requested during call establishment or during the conversation/data phase of the call. When requested during call establishment, the calling customer includes a facility-information element (requesting Service 3) in the set-up message. Upon receipt of such a set-up message, the called customer accepts the invocation in the connect message. Once the call has been established and has entered the conversation/data phase, user-to-user information is exchanged in user-to-user information messages.

If customers wish to invoke Service 3 during the conversation/data phase of the call, and this has not been indicated in the set-up message, then the initiating customer sends a 'facility' message to the other customer with an indication that Service 3 is required. The response is included in a facility message, after which the transfer of information is performed in user-information messages.

9.9 Supplementary services

9.9.1 General

Generic procedures for DSS1 supplementary services are defined by

CCITT [2] but detailed procedures are not given. Three generic protocols are defined to control supplementary services: 'keypad', 'feature-key management' and 'functional'. Keypad and feature-key management protocols adopt a 'stimulus' approach, in which the intelligence of the customer terminal is regarded as low. Terminals using the stimulus approach react to specific instructions and exhibit a low level of processing capability. The functional approach is characterised by requiring a terminal to exhibit a degree of intelligent processing in generating or analysing messages.

9.9.2 Keypad protocol

The keypad protocol is appropriate for basic-rate and primary-rate access. It involves the calling customer requesting supplementary services by using a keypad to generate alpha-numeric codes. The codes are analysed by the network (e.g. in the originating local exchange) to determine the intention of the calling customer. The allocation of codes to supplementary services is not standardised internationally and the codes are therefore network-dependent.

The request to invoke a supplementary service is included in a 'keypad-facility' information element. If the invocation is requested during call establishment, the keypad-facility-information element is included in the set-up message. If the invocation is requested during any other phase of the call, the information element is included in an information message. The network responds to the calling customer in a call-proceeding message (during call establishment) or an information message (during other stages of the call).

To help the calling and called customers understand the progress being made with the supplementary service, information can be supplied to be displayed on the customer terminals. This information is provided by the network in 'display' information elements included in a range of messages.

9.9.3 Feature-key-management protocol

The feature-key-management protocol is applicable to the basic-rate access. It relies upon establishing a 'service profile' for each participating customer. A service profile consists of a file of information which characterises the services offered by the network to that customer. The service profile is agreed between the customer and the network operator upon subscription to the service. The customer requests a supplementary service by sending a 'feature identifier' to the network. The feature identifier is mapped to the corresponding supplementary service by consulting the service profile of the calling customer.

The feature identifier is included in a 'feature-activation' information element, contained within either a set-up message or an information message. The network responds to the customer by including a 'feature indication' information element in a range of call-control messages. Information that can be displayed on the customer's terminal can be sent in a display-information element in a range of call-control messages.

9.9.4 Functional protocol

The functional protocol requires that the customer terminal supporting a supplementary service has a knowledge of that supplementary service. The protocol applies to basic-rate and primary-rate access. There are two categories of functional protocol: 'separate-message' category and 'common-information-element' category. If it is necessary to synchronise the availability of equipment at the customer's premises and at the local exchange, then it is necessary to define specific messages to control supplementary services. These messages belong to the separate-message category. If it is not necessary to synchronise such customer and local exchange resources, then a common-information-element approach can be adopted.

Messages specified in the separate-message category are defined to enact specific functions in the control of supplementary services. The two major functions are 'hold' and 'retrieve'. If a customer sends a hold message to the local exchange during a call, the local exchange reserves the traffic channel (B channel) and the corresponding call reference. A 'hold-acknowledge' message is returned to the customer to confirm that the call has been held. The call is returned to an active state by the customer sending a retrieve message to the local exchange. This is confirmed by the return of a 'retrieve-acknowledge' message.

In the common-information-element category, a 'facility-information' element is used in a generic manner to provide invocation and control of supplementary services. This approach allows new services to be introduced more easily, it supports supplementary services with a large number of variants without a proliferation of messages and it allows multiple-supplementary- service invocations in one message. The facility-information element contains a number of components, including the invoke and return-result components described in Chapter 7 for transaction capabilities. Indeed, the formatting technique adopted in the facility-information element is the same as that described in Chapter 7.

The facility-information element can be applied to supplementary services either involving an existing call or not related to an existing call. For the case when a call exists, or is being established, the facility-information element is included in a facility message or in a set-up message. If a call does not exist, a 'register' message is used to establish an appropriate signalling connection: facility messages can be used once a signalling connection is established.

9.10 Chapter summary

Layer 3 of DSS1 is responsible for setting-up, maintaining and clearing-down circuit-switched calls. It is also responsible for providing access to packet-data facilities.

A Layer 3 message comprises a protocol discriminator, a call reference, a message type and information elements. The protocol discriminator is used to discriminate between circuit-switched call-control messages and other types of message. The call reference is a number that is used to identify the call to which the message pertains. The call reference is chosen by the customer or exchange initiating the call. The message type defines the function of the message being sent. The information elements provide a range of information pertinent to the message e.g. the bearer-capability information element defines a range of features required for a particular call. Information elements can be a single octet in length or of variable length. Each information element is defined by an information-element identifier.

Procedures for setting-up circuit-switched calls can use a broadcast or a point-to-point form of working. In broadcast working, a call is established by sending a set-up message to all terminals associated with the signalling link. All terminals conduct a compatibility check to ensure that communication with the calling terminal can take place. Those terminals that can successfully communicate with the calling terminal respond with an alerting or connect message. The first terminal to respond with a connect message is selected as the receiving terminal. In point-to-point working, an exchange can identify a particular terminal and messages are sent to that terminal.

The clearing of circuit-switched calls is achieved by a three-message sequence using disconnect, release and release-complete messages. Clearing can be initiated by either the calling or the called customer.

Access to packet-data facilities can be circuit-switched or packet-switched. In circuit-switched access, a traffic channel is established to an access unit that provides an interface to a packet-switched public data network. In this case, normal set-up and clear-down procedures apply. In packet-switched access, a packet-handler within an ISDN exchange is used to provide packet-data facilities. In this case, access is gained to the packet handler either by establishing a traffic channel or by using the signalling channel.

User-to-user signalling is provided in three types of service. Service 1 provides an exchange of user-to-user information during the set-up and clearing-down phases of a call. Service 2 allows the transfer of user-to-user information during call establishment between the alerting and connect messages. Service 3 provides for the exchange of user-to-user information during the conversation/data phase of a call.

Detailed procedures for supplementary services are not defined, but three generic procedures are described for DSS1. The keypad protocol is a procedure in which alpha-numeric codes are dialled by the customer, the local exchange analysing the information and acting accordingly. The feature-key-management protocol relies on a service profile being generated for each participating customer, a particular supplementary service being enacted by a customer dialling a code referenced in that customer's service profile. The functional protocol relies on a customer's terminal exhibiting a degree of intelligent processing in generating and analysing messages. In the functional protocol, the terminal is expected to understand a supplementary service and participate in its invocation.

9.11 References

1 CCITT Recommendation Q.931: 'ISDN user–network interface Layer 3 specification for basic call control' (ITU, Geneva)
2 CCITT Recommendation Q.932: 'Generic procedures for the control of ISDN supplementary services' (ITU, Geneva)
3 CCITT Recommendation X.25: 'Interface between data terminal equipment and data circuit terminating equipment for terminals operating in the packet mode' (ITU, Geneva)
4 CCITT Recommendation X.75: 'Packet switched signalling system between public networks providing data transmission services' (ITU, Geneva)

Chapter 10

Interworking of CCS systems

10.1 Introduction

Previous chapters have focused upon the architecture, formats and procedures of the two major CCS systems, viz. CCITT No. 7 and DSS1. There are many principles that are common to the two CCS systems. However, Chapter 1 explains that, because DSS1 is optimised for use in the access network and CCITT No. 7 is optimised for inter-nodal use, the two systems are different when considered at a detailed level. This chapter amalgamates many of the elements that have been covered earlier. Section 10.2 describes the principles adopted by CCITT [1] in specifying the interworking of DSS1 and CCITT No. 7. Section 10.3 then illustrates these principles by giving an example of how DSS1 and CCITT No. 7 interwork to establish and release a basic call in an ISDN. Section 10.4 gives an example of a call that is unsuccessful due to the called customer being busy. More complex examples are given in Sections 10.5 and 10.6 to illustrate the use of non-circuit-related signalling.

10.2 Interworking principles

10.2.1 General

Signalling is the transfer of information between customers, between customers and nodes in the network and between nodes in the network. In this context, customers and nodes can be regarded as 'entities' wishing to communicate. Chapter 4 describes the architecture of modern CCS systems and, in particular, the tiered structure adopted for specification purposes. This structure allows CCS systems to be defined in terms of:

(*a*)   The primitives transferred between adjacent tiers within an entity
(*b*)   The procedures used between corresponding tiers in different entities
(*c*)   The formats of messages exchanged between corresponding tiers in different entities. These constituents are illustrated in Fig. 10.1.

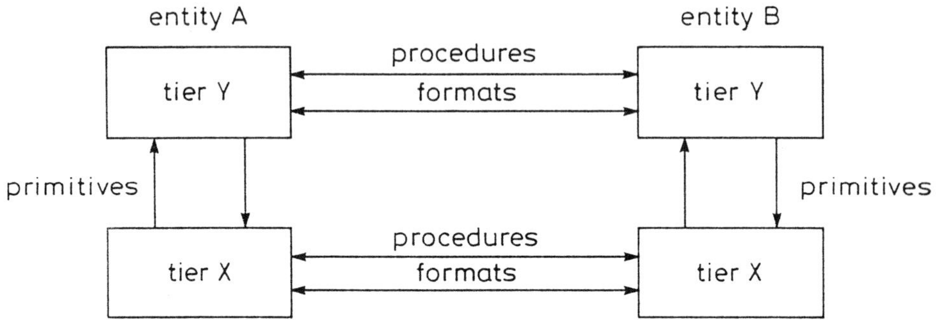

Fig. 10.1   Constituents of interworking

The interworking of DSS1 and CCITT No. 7 takes place at the Layer 3/Level 4 tier. The specification of interworking concentrates on Layer 3 of DSS1 and the ISDN user part (ISUP) of CCITT No. 7. This interworking typically takes place in a local exchange operating in an ISDN. Whilst the constituents of interworking concentrate upon operation within an ISDN, they also have to take into account the need to connect to analogue networks (e.g. an analogue public-switched-telephony network) to complete some calls.

The means of defining the interworking of DSS1 and CCITT No. 7 is to address each of the constituents mentioned above and document the relationship between the two signalling systems.

10.2.2 Primitive constituent

Chapters 1 and 4 describe the close relationship that CCS systems can have with call control, particularly when dealing with circuit-switched calls. The primitive constituent of the interworking definition is covered by considering the relationship between each signalling system and the call control of the local exchange. A model, based upon the reference model for open-systems interconnection [2], is used to apply a disciplined approach. The model is illustrated in Fig. 10.2 for the case of a customer originating a call.

The model consists of three groups of functions: call control, DSS1 Layer 3 and CCITT No. 7 ISUP. The call-control group of functions acts as an intermediary between the two signalling systems. Each signalling system communicates with the call control by means of primitives. When a signalling system transfers a primitive to the call control, call processing within the call control takes place; this results in a primitive being returned to the same signalling system or to the other signalling system. There are four types of primitive. The 'indication' primitive is issued by a signalling system to invoke call-control processing. The 'response' primitive is issued by call control to signify the completion of call processing initiated by an indication primitive. The 'request' primitive is issued by call control to invoke a signalling procedure. The 'confirm' primitive is issued by a signalling system to signify completion of a signalling procedure initiated by a request primitive.

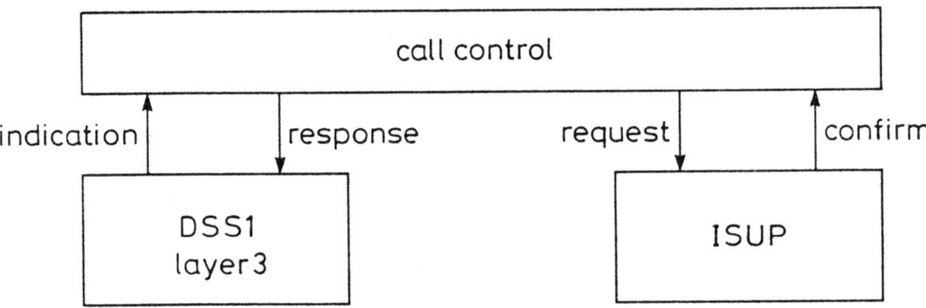

Fig. 10.2   Primitive constituent model

To avoid confusion over which functions form part of call control and which functions are part of the signalling system, it is necessary to distinguish between the primitive-constituent model for interworking and practical implementations in networks. In the primitive-constituent model, call processing (i.e. analysis and decision-making) is assumed to take place in the call control. This is assumed so that a disciplined approach can be adopted to interworking, ensuring that a comprehensive specification results. In practical implementations, functions are not necessarily carried out in the structure that is assumed for modelling purposes. Thus, processing can occur in call control, in the signalling system itself, or even in an element of equipment that does not fit into the structure assumed for specification purposes. Hence, whilst the primitive-constituent model is very useful in defining the interworking requirements of DSS1 and ISUP, it should not be assumed that practical implementations adopt the structure of the model.

10.2.3 Procedure constituent

Procedures can be documented by means of time-sequence diagrams or the 'specification and description language (SDL)' [3]. Time-sequence diagrams are similar to those used in previous chapters of this book, the diagrams showing the flow of messages in a logical sequence. The diagrams are supported by descriptive text. The SDL is a system specified by CCITT that documents the status of an interface using symbols to denote incoming and outgoing messages and call-processing decisions. The SDL is a very-effective means of specifying procedures, particularly under failure conditions. However, this book continues to use the time-sequence diagram as the most-effective means of explaining message flows. The general format of a time-sequence diagram used for interworking is given in Fig. 10.3. Message A is sent from the customer's DSS1 Layer 3 to the corresponding function at the local exchange. This results in a primitive being sent to the local-exchange call control. The call control processes the information provided by the primitive and decides on an appropriate action to take. In this case, the appropriate action is determined to be to send a message to another exchange. Hence, a primitive is sent to the local-exchange ISUP requesting a message to be formulated and routed accordingly. The ISUP thus sends Message B to the next exchange involved in the call. Similar activities apply to Messages C and D. Although primitives are generated within the local exchange, the procedure constituent sees only an exchange of messages in a time sequence.

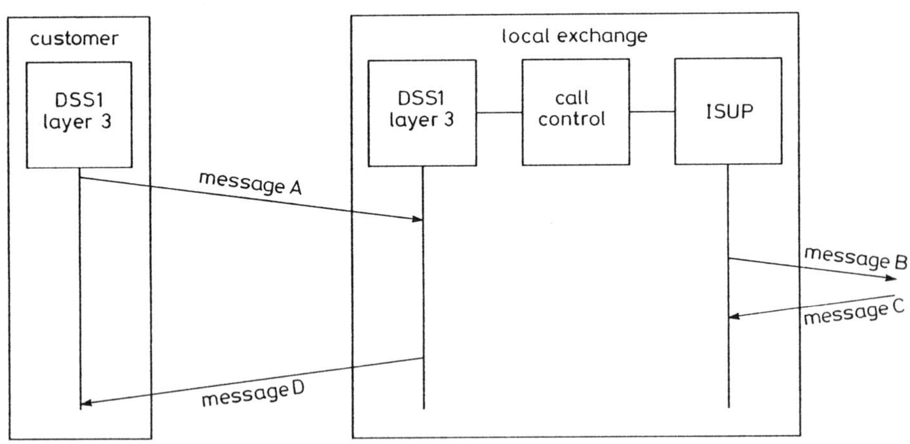

Fig. 10.3   General format of time-sequence diagram

10.2.4 Format constituent

The type of message and the content of each message in DSS1 Layer 3 is compared with the equivalent information for the ISUP in a 'mapping table'. Thus, the information elements of DSS1 are mapped into the parameters of the ISUP. In some cases, there is a direct correlation between an information element and a parameter, whereas in other cases only a subset of an information element is mapped into a parameter. Information elements of DSS1 that are of local significance (i.e. relevant only to the access link) are not considered.

10.3 Example of interworking for a basic call

10.3.1 General

Consider the case of Customer A (connected to Local Exchange A) initiating a call to Customer B (connected to Local Exchange B). Both customers are connected to respective local exchanges by digital transmission links and the terminals at the customers' premises both operate DSS1. The network connecting the two customers is an ISDN. The establishment and release of the call can be described in terms of the three constituents. The procedure constituent is considered in Section 10.3.2, the primitive constituent is covered in Section 10.3.3 and the format constituent is illustrated in Section 10.3.4.

10.3.2 Procedure constituent

The procedure constituent for basic-call interworking of DSS1 and ISUP is illustrated in Fig. 10.4. The entities involved in the call are Customers A and B, respective local exchanges and a trunk exchange. In this example, both customers operate the en-bloc form of working and Customer B has a non-automatic-answering terminal.

Customer A initiates the call by instructing the terminal to establish a call to Customer B and providing the appropriate address of Customer B. If the call is a telephony call, this is equivalent to Customer A picking up a telephone handset and dialling (or keying) the telephone number of Customer B. These activities result in a set-up message being sent to Local Exchange A. The set-up message includes the address of Customer B and the type of connection required. Local Exchange A analyses the set-up message and recognises that the call needs to be routed via the trunk exchange. Thus, the ISUP of Local Exchange A formulates an initial-address message (IAM) and sends it to the trunk exchange. Local Exchange A also returns a call-proceeding message to Customer A to indicate that the call is being processed.

Upon receipt of the IAM, the trunk exchange analyses the address of Customer B and determines that the call needs to be routed to Local Exchange B. Thus, the trunk exchange formulates an appropriate IAM and sends it to Local Exchange B. Local Exchange B analyses the information contained within the IAM and determines the identity of Customer B. A set-up message is sent to Customer B and an address-complete message is returned to the trunk exchange to indicate that enough information has been received to identify Customer B. In this example, Local Exchange B recognises that Customer B does not have a multi-terminal configuration requiring the broadcast form of working. Thus, the point-to-point form of working is adopted.

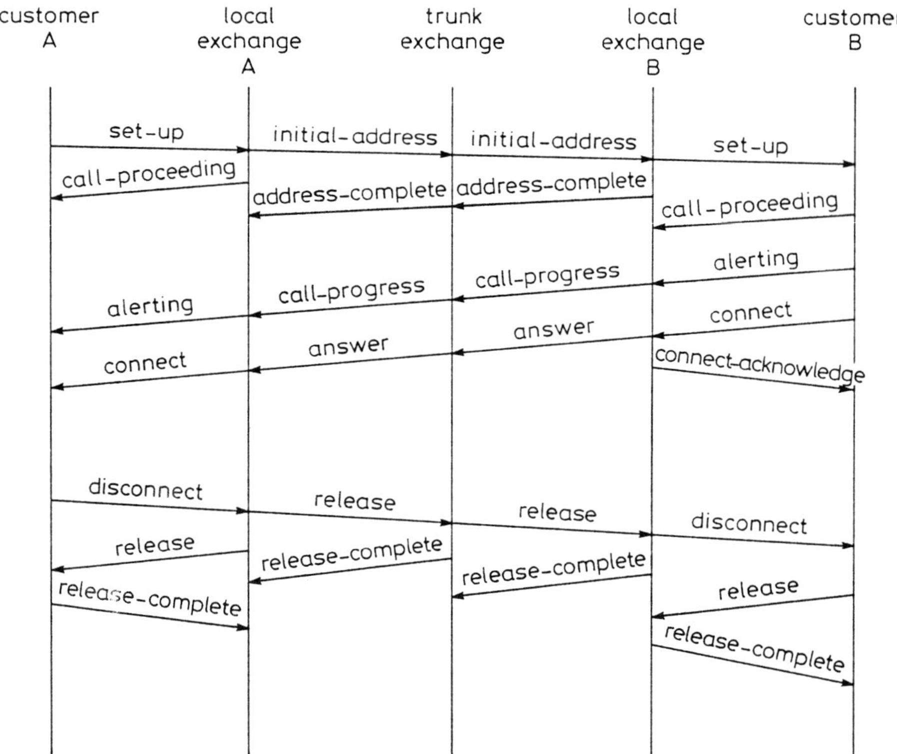

Fig. 10.4   Procedure constituent for basic-call interworking

Upon receipt of the set-up message, the Customer B terminal adopts the option of returning a call-proceeding message to Local Exchange B to indicate that the call is being processed. This message does not involve actions at Local Exchange B, other than to re-set internal timers. When the Customer-B terminal begins to alert Customer B that an incoming call is being made (e.g. a telephone begins to ring), an alerting message is returned to Local Exchange B. Local Exchange B recognises that alerting has commenced and sends a call-progress message to the trunk exchange; this, in turn, sends a call-progress message to Local Exchange A. Local Exchange A informs Customer A of the status of the call by sending an alerting message.

When the call is answered by Customer B (e.g. the telephone handset is lifted), a connect message is sent to Local Exchange B. Local Exchange B returns a connect-acknowledge message to Customer B and sends an answer message to the trunk exchange. The trunk exchange passes an answer message to Local Exchange A, which completes the set-up sequence by sending a connect message to Customer A indicating that the call is established. In this example, the option for Customer A to send a connect-acknowledge message to Local Exchange A is not adopted.

The call can be cleared-down by either customer: in this example, Customer A initiates the release sequence by sending a disconnect message to Local Exchange A. This results in Local Exchange A sending a release message to the trunk exchange and a release message to Customer A. Customer A responds with a release-complete message. Upon receipt of a release message, the trunk exchange sends a release message to Local Exchange B and a release-complete message to Local Exchange A. Upon receipt of a release message at Local Exchange B, a disconnect message is sent to Customer B and a release-complete message to the trunk exchange. Finally, upon receipt of a release message from Customer B, Local Exchange B sends a release-complete message to Customer B.

10.3.3 Primitive constituent

In this constituent, each exchange has an 'incoming' signalling system (defined as receiving a set-up or initial-address message), an 'outgoing' signalling system (defined as sending a set-up or initial-address message) and a call control. Customer A has an outgoing DSS1 and Customer B has an incoming DSS1. For explanation purposes, consider the portion of the call involving Customer A and Local Exchange A, as shown in Fig. 10.5. Similar explanations apply to the portions of the call involving the trunk exchange, Local Exchange B and Customer B.

Customer A initiates the call, resulting in a set-up-request primitive being sent to the outgoing DSS1 of Customer A. The outgoing DSS1 formulates a set-up message, including the address of Customer B and the type of connection required. The set-up message is sent to the incoming DSS1 of Local Exchange A, which causes a set-up-indication primitive to be sent to the call control of Local Exchange A. The call control analyses the information provided in the primitive and takes three courses of action. First, it returns a proceeding -request primitive to the incoming DSS1, thus causing a call-proceeding message to be returned to Customer A. Secondly, the call control recognises that a call needs to be established via a trunk

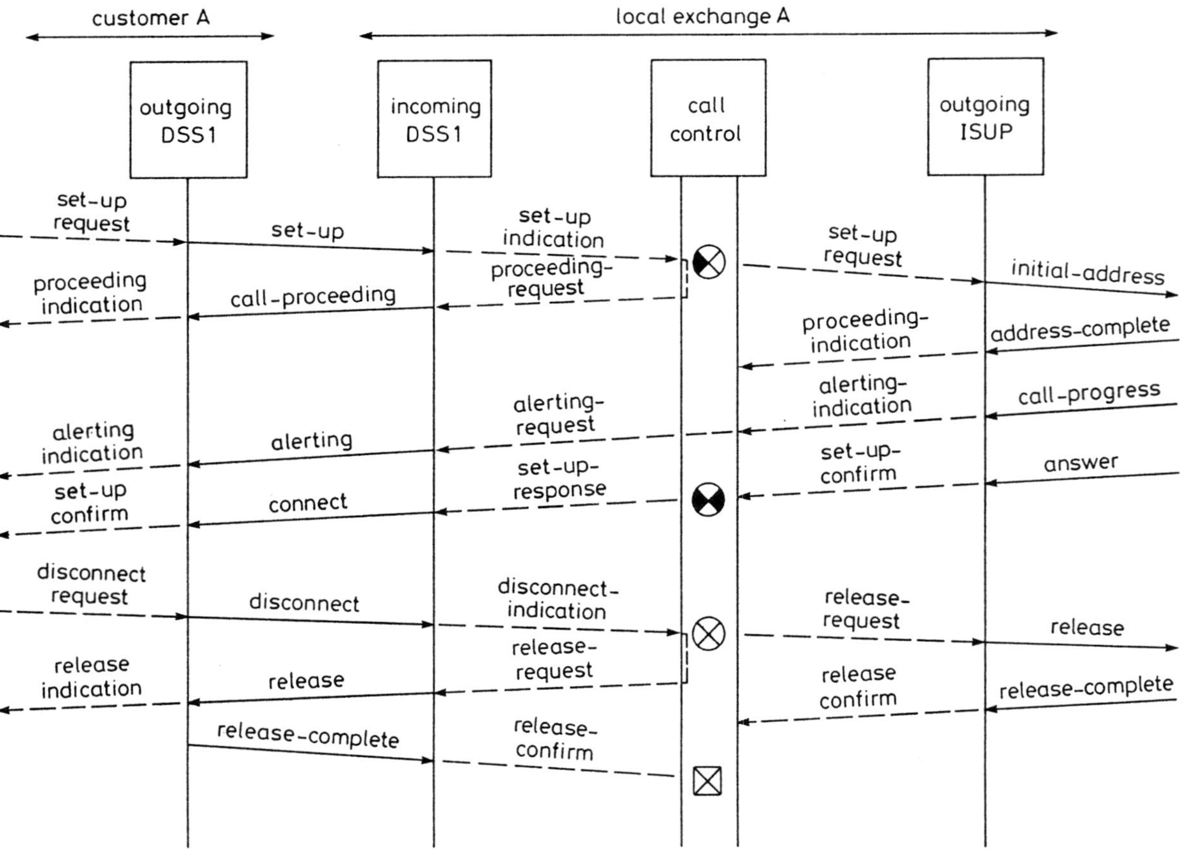

Fig. 10.5   Primitive constituent for basic-call interworking

exchange and it requests the outgoing ISUP to formulate an initial-address message (IAM) by sending a set-up-request primitive. The outgoing ISUP acts upon the request by formulating an IAM and sending it to the appropriate trunk exchange. The third action of call control is to instruct the switch block to connect-through the traffic circuit involved in the call in the backward direction (so that Customer A can hear any tones provided by the network, if provided).

When the outgoing ISUP receives an address-complete message from the trunk exchange, a proceeding-indication primitive is sent to the call control. Receipt of this primitive does not result in the call control issuing a corresponding primitive; it allows the call control to release certain specialised information that is held in short-term memory for the call. For example, specialised routeing information used to set-up the call can be released when it is clear that sufficient information has been received at Local Exchange B to identify Customer B.

The next message to be received by the outgoing ISUP is call progress, indicating that Customer B is being alerted. This results in an alerting-indication primitive being passed to the call control. The call control recognises that Customer B is being alerted and that Customer A should be informed of the status of the call. An alerting-request primitive is sent to the incoming DSS1, resulting in an alerting message being sent to Customer A.

When Customer B answers the call, an answer message is returned through the network to the outgoing ISUP at Local Exchange A. This results in a set-up-confirm primitive being passed to the call control. The call control recognises that Customer B has answered the call and instructs the switch-block to connect-through the traffic circuit in the forward direction; this completes the traffic-circuit connection. The call control also sends a set-up-response primitive to the incoming DSS1, causing the transmission of a connect message to Customer A. The connect message indicates that the set-up sequence is complete.

The release of the call is initiated by Customer A. This results in a disconnect-request primitive being sent to the outgoing DSS1 at Customer A. This, in turn, results in a disconnect message being sent to the incoming DSS1 at Local Exchange A. The receipt of the disconnect message causes a disconnect-indication primitive to be sent to the call control.

Upon receipt of the disconnect-indication primitive, the call control performs three activities. First, the call control recognises that the call needs to be cleared down in the network. Thus, the call control issues a release-request primitive to the outgoing ISUP, causing the sending of a release message to the trunk exchange. Secondly, the call control instructs the switch-block to release the speech circuit. Thirdly, the call control recognises that the access link needs to be cleared-down. It issues a release-request primitive to the incoming DSS1, thus causing a release message to be sent to Customer A. When the release sequence is completed, the call control receives a release-confirm primitive from the incoming DSS1 and outgoing ISUP. Upon receipt of the release-confirm primitive from the incoming DSS1, the call control recognises that there is no further use for the access call reference and the final act is to release it and return it to the common pool, ready for allocation to another call.

10.3.4 Format constituent

In this constituent, the DSS1 messages used in the procedure constituent are mapped onto the ISUP messages. The constituent thus applies to the two points of interworking, i.e. the two local exchanges. The mapping is presented in terms of the messages sent from Customer A to the network and from the network to Customer B. The mappings given below are illustrative and are not intended to be exhaustive, e.g. only some of the optional information elements and parameters are described.

The set-up message is mapped onto the initial-address message, as shown in Table 10.1. The bearer-capability-information element defines a large range of features relating to the type of connection and service required by Customer A. The bearer-capability-information element is mapped unchanged into an optional IAM parameter called 'user-service information'. This parameter is transported through the network to Local Exchange B, thus allowing it to be mapped back into the bearer-capability-information element in the set-up message to Customer B. However, a sub-set of the information in the bearer-capability-information element is required to enable the correct type of connection to be chosen when routeing the call through the network. The sub-set chosen is incorporated into the 'transmission-medium-requirements' parameter. For example, if it is necessary that a 64 kbit/s traffic circuit is provided between Customers A and B, then the transmission-medium-requirements parameter in the IAM ensures that such a circuit is selected as the call is routed through the network. The mapping described allows the network to transport transparently (i.e. without analysis) the user-service-information parameter whilst using a sub-set (in the form of the transmission-medium-requirements parameter) to route the call.

The forward-call-indicator parameter is used in the IAM to indicate the signalling capability of the call being established, e.g. whether or not end-to-end signalling is possible. This is an example of a parameter that is relevant to the network but for which there is not a mapping to a DSS1 information element.

The progress-indicator-information element is used to keep the customers informed of events that occur during the call. This information is not directly relevant to the network. It is therefore carried in a parameter called 'access transport', which is transferred between the local exchanges without being analysed.

Table 10.1 Mapping of set-up and initial-address messages

Customer A ⟶ Set-up message	Network ⟶ Initial-address message	Customer B Set-up message
Information element	Parameter	Information element
Bearer capability	User-service information Transmission-medium requirements	Bearer capability
No mapping	Forward-call indicator	No mapping
Progress indicator	Access transport	Progress indicator
Calling-party number	Calling-party number	Calling-party number
Called-party number	Called-party number	Called-party number

Table 10.2 Mapping of connect and answer messages

Customer A ⟵ Connect	Network ⟵ Answer	Customer B Connect
Information element	Parameter	Information element
Progress indicator	Access transport	Progress indicator

Table 10.3 Mapping of disconnect and release messages

Customer A ⟶ Disconnect	Network ⟶ Release	Customer B Disconnect
Information element	Parameter	Information element
Cause	Cause	Cause

The calling-party number and called-party number are examples of information elements that are mapped directly into IAM parameters.

There is no mapping for the call-proceeding message because it is of local significance, i.e. it applies only to the access links. The mapping for the connect and answer messages is given in Table 10.2. The explanation of the mapping of the progress-indicator-information element and the access-transport parameter is given above. The mapping of the disconnect and release messages is given in Table 10.3. The 'cause-information' element indicates the reason for sending the disconnect or release message. In the example of Customers A and B, the cause is coded as 'normal call clearing'.

10.4 Example of interworking for an unsuccessful basic call

It is useful to consider an example of an unsuccessful call but, for brevity, only the procedure constituent of interworking is described. Consider the case described in Section 10.3 of Customer A attempting to establish a call to Customer B. However, in this example, Customer B is already busy with another call and cannot accept the call attempt by Customer A. The procedures are illustrated in Fig. 10.6.

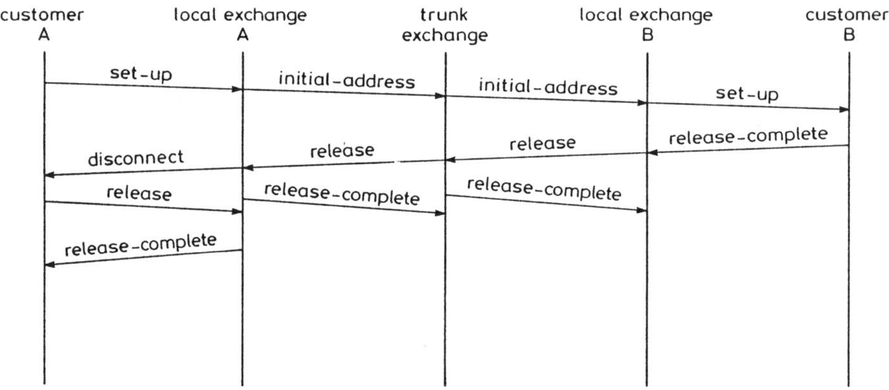

Fig. 10.6   Procedure constituent for unsuccessful call

The call is initiated by Customer A sending a set-up message to Local Exchange A. This causes the sending of an IAM through the network. Local Exchange B sends a set-up message to Customer B. However, Customer B is busy. Hence, instead of responding with a call-proceeding, alerting or connect message, Customer B sends a release-complete message to Local

Exchange B. The release-complete message contains an information element termed 'user busy' to indicate the reason for refusing the call. The release-complete message is mapped into a release message to initiate the normal clear-down procedure in the network. The release message includes a cause parameter coded as user busy. Upon receipt of a release message at Local Exchange A, the normal release procedures are enacted. The disconnect message includes the cause-information element to inform Customer A of the reason for the call being unsuccessful.

10.5 Example of database-access call

Exchanges within a network store a great deal of information to allow calls to be routed on demand. However, some routeing information is specialised in nature and it can be more economic to store such information in a database rather than duplicate it in a large number of exchanges. In this case, exchanges that are nominated to gain access to specialised routeing information need to recognise that a particular call requires the interrogation of a database.

A typical network structure, in which a database contains specialised routeing information, is illustrated in Fig. 10.7. The normal routeing of a call involves the originating local exchange, the trunk exchange and the terminating local exchange. However, in this case, the trunk exchange recognises that specialised routeing information is required to complete the call and that it is stored in a database. The trunk exchange thus gains access to the database, to determine the routeing information. Using the specialised routeing information, the trunk exchange completes the call by routeing it to the appropriate terminating local exchange.

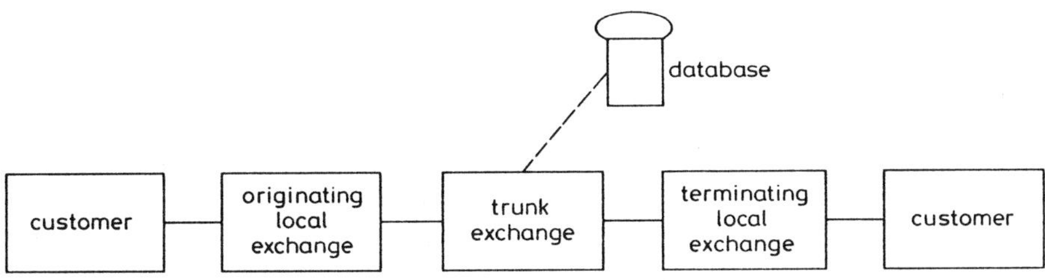

Fig. 10.7   Typical network structure

An example of the database approach is when a network operator implements a service in which a call is charged to the called customer rather than the calling customer. This service is often called 'freephone' or 'toll-free' because each call is free to the calling customer. An effective way of implementing such a service [4,5,6] is to dedicate a 'national-number-group (NNG) code' to the service. A NNG code is the number that is normally associated with a geographic area when dialling a trunk (or long-distance) call. A typical value of NNG for toll-free service is 800. Thus, a calling customer typically dials a code of +800XXXXXX, where the initial + is a prefix associated with a trunk or long-distance call (typically 0 or 1), the 800 code identifies the toll-free service and the remainder of the number identifies the called customer. In the structure shown in Figure 10.7, the originating local exchange only needs to recognise the prefix or the 800 NNG and pass the call to the trunk exchange. The trunk exchange recognises the 800 NNG and initiates an interrogation of the database to determine the routeing of the call.

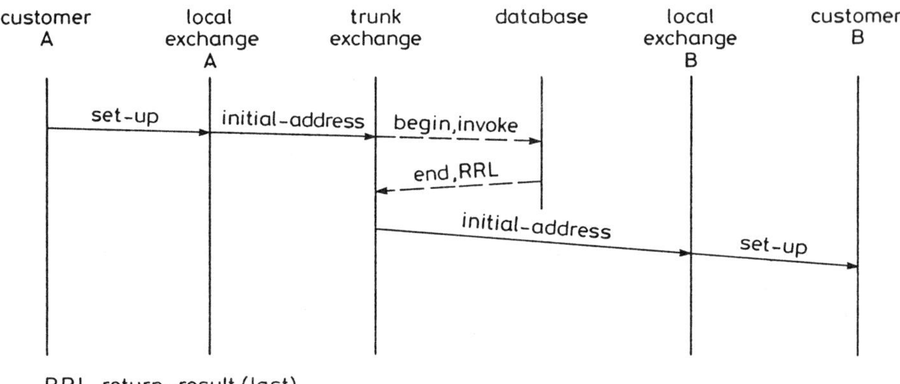

Fig. 10.8   Procedure constituent for simple database call

Consider a customer initiating a freephone call, when the service is implemented according to the network structure described above. The procedure constituent is illustrated in Fig. 10.8. Customer A dials (or keys) the number +800XXXXXX. The customer terminal formulates a DSS1 set-up message with the dialled number as the called-party-number-information element. This message is sent to Local Exchange A. Local Exchange A analyses the set-up message and, identifying either the prefix or the 800 NNG code, recognises that the call needs to be passed to the trunk exchange. Thus, the Local Exchange A ISUP formulates an initial-address message (IAM), including a parameter containing the called-party number, and sends it to the trunk exchange. Local Exchange A also establishes a traffic circuit to Customer A and to the trunk exchange.

Upon receipt of the IAM, the trunk exchange recognises that the call is a freephone call and that specialised routeing information is required from the database before the call can be established. The trunk exchange therefore establishes a communication with the database. In this communication, there is no need for a traffic circuit: only an exchange of signalling information is required. The non-circuit-related form of CCITT No. 7 (Transaction Capabilities) is therefore used to conduct the communication with the database. The trunk exchange formulates a begin message, to establish the communication, including an invoke component. The format of a typical begin message is given in Fig. 10.9. The begin message includes a transaction identity that is used to correlate the messages used in the communication with the call. The invoke component contains an invoke identity that is used to correlate the response to the invoke component. The invoke component also includes an operation code that requests the database to examine the called-party number and provide routeing information. The called-party number (i.e. the value received by the trunk exchange in the incoming IAM) is included in the parameter field following the operation code.

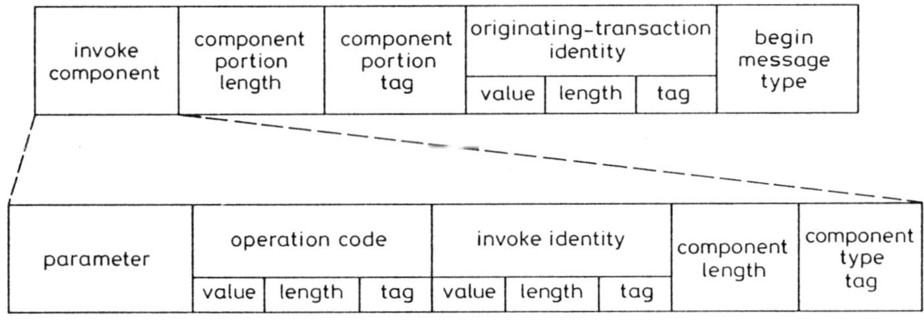

Fig. 10.9   Format of typical begin message

Upon receipt of the begin message, the database analyses the invoke component to determine the nature of the request. It recognises that the operation code requests routeing information for the called-party number and it consults its records to provide the required information. When the routeing information has been retrieved, it is included in an end message with a return-result (last) component. The end message is used to signify that the communication is completed. In this example, the routeing information is supplied in the form of a new called-party number, with the 800 NNG code being translated into a traditional geographically-based NNG and the XXXXXX being translated into a new customer number.

When the trunk exchange receives the end message, it extracts the new called-party number and recognises that the communication with the database is now completed for this call. The trunk exchange analyses the location of Customer B and formulates an ISUP IAM including the new called-party number. The IAM is sent to Local Exchange B and a traffic circuit is also established to Local Exchange B. As far as Local Exchange B is concerned, the incoming call is a normal circuit-related call and it is treated accordingly to complete the set-up sequence.

10.6 Example of a complex database-access call

In this example, suppose that the same network structure exists as that described in Section 10.5. However, in this case, the sequence is made more complex by the database needing additional information about Customer A before it is able to supply the routeing information to the trunk exchange. For example, Customer B has two locations, one dealing with enquiries from

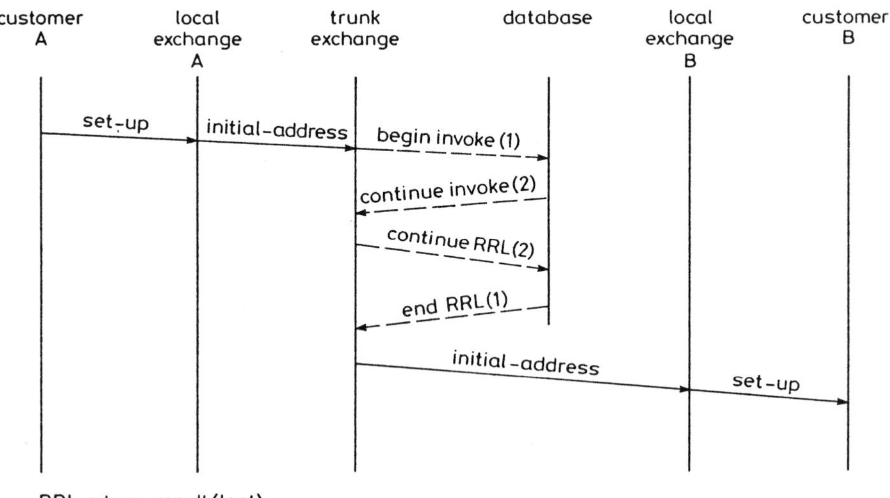

Fig. 10.10   Procedure constituent for complex database call

customers in one area and the other dealing with enquiries from another area. In this example, the set-up sequence is the same as that described in Section 10.5 up to and including the trunk exchange establishing a communication with the database. The additional sequences are shown in Fig. 10.10. A begin message with an invoke component (Invoke 1 in Fig. 10.10) is sent to the database. In this case, the database needs to know to which location to route the call. To ascertain this additional information, the database issues a continue message that includes an invoke component (Invoke 2). The continue message is used to indicate that the message refers to the communication under consideration. The invoke component (Invoke 2) includes an operation code that requests the trunk exchange to supply the calling-party number. Upon receipt of the continue message, the trunk exchange analyses the request for the calling-party number and supplies it in a continue message that includes a return-result (last) component (2). The return-result (last) component refers to the invoke component sent by the database (Invoke 2).

Receipt of the calling-party number allows the database to select the appropriate location of the Customer B and supply the corresponding called-party number in an end message with a return-result (last) component (1). This return-result (last) component refers to the invoke component sent by the trunk exchange (Invoke 1). The call set-up process continues as described in Section 10.5.

10.7 Chapter summary

Modern CCS systems are specified in terms of:

(a)   The primitives transferred between adjacent tiers within an entity
(b)   The procedures used between corresponding tiers in different entities
(c)   The formats of messages exchanged between corresponding tiers in different entities.

These constituents are used to define the interworking of CCS systems. For DSS1 and CCITT No. 7, the interworking takes place at the Layer 3/Level 4 tier.

For the primitive constituent, a model has been devised comprising call control, DSS1 and CCITT No. 7. The call control acts as a processor, taking inputs from the signalling systems in the form of primitives, deciding upon action to be taken and issuing responses to the signalling systems. All communication between call control and the signalling systems in the model is conducted using the indication, response, request and confirm primitives. The model provides a disciplined approach to interworking but it does not reflect practical implementations, e.g. in practice, some of the call processing can take place within the signalling systems.

The procedure constituent uses time-sequence diagrams similar to those used in earlier chapters. Message flows into and from entities are defined.

These message flows illustrate the logical sequence of events that can take place for various types of call.

The format constituent is defined by mapping tables. Each DSS1 message is mapped onto a CCITT No. 7 message, unless a particular message is of local significance only. Similarly, information elements of DSS1 are mapped onto parameters of CCITT No. 7.

Examples of the interworking of DSS1 and CCITT No. 7 are given for a basic call, an unsuccessful call and calls needing the support of remote-network databases.

10.8 References

1 CCITT Recommendation Q.699: 'Interworking Between Digital Sub-scriber Signalling System No. 1 and Signalling System No 7' (ITU, Geneva)
2 CCITT Recommendation X.200: 'Data communications networks open systems interconnection model and notation service definition' (ITU, Geneva)
3 CCITT Recommendation Z.100: 'Functional specification and description language criteria for formal description techniques' (ITU, Geneva)
4 Manterfield, R J: 'Specification and evolution of CCITT No. 7', Proceedings International Switching Symposium, Phoenix, 1987
5 Manterfield, R J: 'Migration of intelligence in evolving Networks', Forum 1987, ITU Technical Symposium, Geneva
6 Ward and Lark: 'The DDSN on line in the United Kingdom', Proceedings International Switching Symposium, 1990

Chapter 11

Conclusions

A telecommunications network is implemented to provide services to customers. The objective is to provide features that quickly meet the needs of customers to a high level of quality and at low cost. The basic foundation of a telecommunications network comprises transmission systems and exchanges. Transmission systems provide the ability to transfer speech and data. Exchanges are used to switch transmission links for calls, thus allowing customers to share transmission links. Signalling provides the ability to transfer information between customers and networks, within networks and between customers. Signalling is the life-blood, the vitalising influence, of a network. Signalling information flows through a network to transform it from an inert aggregate of elements to a powerful medium providing services to customers. Signalling is the bond that makes a network a cohesive force.

Signalling systems can be 'channel associated' or 'common channel'. In channel-associated signalling (CAS) systems, signalling capacity is dedicated for each traffic circuit. CAS systems are optimised for use in old-technology networks (e.g. using electromechanical exchanges). In common-channel signalling (CCS) systems, signalling capacity is provided in a common pool and the capacity is allocated dynamically as required. CCS systems are optimised for use in modern-technology networks (using software-controlled exchanges). This book has concentrated on describing modern CCS systems.

In CCS systems, signalling information is transferred in messages. Error-control mechanisms are used to ensure that the messages are transferred without corruption. Early CCS systems (e.g. CCITT Signalling System No. 6) were 'circuit-related' in nature, meaning that each message is identified by the number of the traffic circuit to which the message refers. Modern CCS systems (e.g. CCITT Signalling System No. 7) exhibit both circuit-related and 'non-circuit-related' attributes. In non-circuit-related cases, a message is identified by a reference number, rather than a traffic-circuit identity, thus allowing signalling to take place in the absence of a related traffic circuit. The major advantages of modern-CCS systems are that they:

(*a*)   exhibit a large degree of evolutionary potential,
(*b*)   have a wide repertoire of signals,

(*c*)   are compatible with software-controlled exchanges and customer terminals, and

(*d*)   form comprehensive signalling networks allowing increased security and advanced operational, maintenance and administration facilities.

In the past, CCITT concentrated on the specification of inter-exchange CCS systems. The first CCS system to be specified internationally was CCITT No. 6. This signalling system was implemented widely, but it is now being superseded by CCITT No. 7. CCITT No. 7 exhibits more evolutionary potential, has a flexible coding format, can handle non-circuit-related applications and is optimised for a digital environment. As ISDNs are being conceived and implemented, it is recognised that customers can only achieve the full benefits of this modern technology if modern CCS systems are provided between customers and their respective local exchanges. Thus, CCITT has specified the Digital Subscriber Signalling System No. 1 (DSS1).

Modern CCS systems are extremely complex and it is essential that there is a disciplined and logical approach to their specification. The structure of the specification is called its 'architecture'. Both CCITT No. 7 and DSS1 follow a rigorous approach in which they are specified in a number of tiers. The circuit-related aspects of CCITT No. 7 are defined in four tiers called 'levels'. Levels 1 to 3 constitute the message-transfer part (MTP) which is responsible for transferring messages between nodes in the network. Message transfer occurs without loss or corruption and the messages are delivered in the same order as they are transmitted, even when there are failures in the network. Level 4 contains the user parts of CCITT No. 7, which define the meanings of the messages and the logical sequence of events that can occur for specific applications (e.g. telephony). Three user parts are specified: the telephone user part (TUP), ISDN user part (ISUP) and data user part (DUP).

The non-circuit-related aspects of CCITT No. 7 are defined in accordance with the open-systems interconnection (OSI) 7-layer model. Layers 1 to 3 comprise the combination of the MTP and the 'signalling-connection-control part (SCCP)'. This combination offers a network service to higher-layer functions. Layers 4 to 7 contain 'transaction capabilities (TC)'. TC is a protocol that is application-independent, i.e. it can be used by a range of applications to transfer non-circuit-related information between network nodes.

DSS1 is defined in the context of the OSI 7-layer model and comprises three layers. Layer 1 defines the electrical, physical and functional characteristics of a customer's line. Layer 2 is responsible for transferring messages between a customer and a local exchange. Layer 3 defines the meaning of the messages being transferred and the logical sequence of events for ISDN calls.

Nodes in the network and the terminal equipment of customers can be termed 'entities' for specification purposes. The specifications of modern CCS systems are defined in terms of:

(i)   the primitives transmitted between adjacent tiers within an entity,
(ii)   the procedures adopted between corresponding tiers in different entities and
(iii)   the format of the messages used between corresponding tiers in different entities.

The interworking of CCS systems is also specified in such terms. For DSS1 and CCITT No. 7, the interworking is defined at the Layer 3/Level 4 tier.

The international specification process conducted by CCITT results in recommendations that define CCS systems. The process is invaluable in that it allows the ideas and views of experts throughout the world to be harnessed in a common goal. The recommendations provide standards that are of very high quality and comprehensive, taking account of failure modes as well as successful calls. These standards provide a common reference to which implementations can be designed. This allows the development of CCS systems to benefit from the economies of scale of all exchange manufacturers working to a common standard. Despite this process, it is still difficult to produce specifications that are completely unambiguous. Feedback must be used, from trials of early systems and implementations, to improve continuously the quality of the specifications.

In the past, the specification process has been slow and incomplete. The slow specification of a new idea, or method of achieving an objective, places network operators in a dilemma. If customers need a new facility, network operators can either wait for a standard method of implementation to be specified or they can implement quickly their view of the expected standard. If they wait, they do not fulfil the needs of the customer. If they implement quickly, the resulting interim systems rarely match the final version of the standard. Thus, network operators can bear the burden of developing equipment that does not benefit from the economies of scale of standard equipment. They can also be left with non-standard equipment that incurs extra costs when it is necessary to evolve.

An incomplete specification process is one in which there is not full agreement on the way forward. Rather than resolve to achieve one method of performing a function, several methods are specified or options are adopted. These extra methods and options complicate the specification and increase the cost of implementation. Interestingly, a slow specification process can exacerbate this problem because, once network operators have implemented an interim system, they can be reluctant to agree new standards that make evolution difficult. Instead of all participants starting from a 'common platform with a common goal, vested interests can cause obstacles to progress.

A solution to the problem is to speed-up the specification process. Once an idea or new method is considered viable and desirable, a concerted effort should be made to specify it before interim systems, exhibiting limited evolutionary capability, are adopted. More examination of proposals to define options should also be performed, thus reducing the range of national variants. The production of effective standards quickly will reduce the number of variants adopted, thus decreasing costs of implementation, whilst providing a greater range of services to customers.